Pyrolysis—Gas Chromatography

Analytical Sciences Monographs

Pyrolysis—Gas Chromatography

R. W. May

Home Office Forensic Science Laboratory
Aldermaston, Reading, Berkshire, RG7 4PN

E. F. Pearson
& D. Scothern

Home Office Central Research Establishment
Aldermaston, Reading, Berkshire, RG7 4PN

The Chemical Society
Burlington House
London W1V 0BN   1977

© The Chemical Society 1977
ISBN: 0 85186 767 7
ISSN: 0583-8894

Printed by Heffers Printers Ltd Cambridge England

Contents

Preface

Many papers have been published, particularly over the past decade, on aspects of pyrolysis–gas chromatography. A large number of different types of apparatus have been used, on a wide range of samples. This monograph attempts to present the available knowledge in a form useful to the practising analyst, helping in the choice of an appropriate method and in the avoidance of the more common pitfalls in this, perhaps deceptively, simple technique.

Chapter 1 serves as an introduction to gas chromatography and will be of interest to those unfamiliar with the technique. The several methods of pyrolysis used in pyrolysis–gas chromatography are described in Chapter 2; their merits and demerits in particular applications are discussed. The major analytical uses of the technique are presented in Chapter 3; the general analytical 'fingerprinting' aspects are described separately from the method as applied to specific sample types. Chapter 4 deals with the identification of the pyrolysis products which are eluted from the chromatograph column, useful extra information allowing the possibility of naming a pyrolysed sample without recourse to a known identical sample. The necessity for increased standardization of the technique of pyrolysis–gas chromatography is discussed in Chapter 5, and the method used in these laboratories is detailed.

We should like to take this opportunity to thank the following persons who have contributed to the completion of this monograph; Dr A. S. Curry, Controller, Forensic Science Service, P. G. W. Cobb, director, Forensic Science Laboratory, Harrogate, B. German, who carried out the mass spectrometry, and J. Porter, who carried out the major part of the experimental work.

R. W. May
E. F. Pearson
D. Scothern

1 Gas Chromatography

History

The word 'chromatography' was first coined in 1906 by Tswett[1] to describe his technique for separating the components of plant pigments by introducing the mixture onto the top of a column of solid adsorbent and allowing solvent to percolate down the column. He found that the different components were carried down the column at different rates, and thus became separated, to form discrete coloured bands. The term is now used to describe any technique where a separation of the components of a mixture is achieved by their distribution between a fluid mobile phase and a stationary phase.

Tswett's work was an example of what came to be known as 'liquid–solid chromatography', as the mobile phase was a liquid and the stationary phase a solid. It can also be classified as 'adsorption chromatography', in that the essential mechanism involves the adsorption of the pigments on the solid stationary phase.

In 1941, Martin and Synge[2] described 'partition chromatography'. They used silica gel impregnated with water as the stationary phase, and a water-immiscible liquid as the mobile phase. The silica gel acts merely as a support for the water, and under ideal conditions does not influence the separation, which is achieved by a liquid–liquid partition between the mobile phase and the water. In the same paper the authors also suggested that the mobile phase need not be a liquid, but may be a vapour, and they showed that the efficiency of contact between the phases (theoretical plates per unit length of column) is far greater in the chromatogram than in ordinary distillation or extraction columns. They claimed that very refined separations of volatile substances should therefore be possible by a column in which a permanent gas is made to flow over gel that is impregnated with a nonvolatile solvent in which the substances to be separated approximately obey Raoult's Law; when differences of volatility are too small to permit ready separation by these means, advantage may be taken in some cases of deviations from Raoult's Law, as in azeotropic distillation.

This remarkable prediction has now been borne out in every detail, but no one took it up until Martin and James began the development of gas–liquid chromatography a number of years later, and in a series of papers, in 1952, Martin and James laid the groundwork to the subject.[3-6] These early separations were of a series of organic acids and of organic bases, using acid–base titration to detect and measure the components as they emerged from the end of the column. Figure 1.1 shows one of their separations of the fatty acids up to valeric acid, on a column 11 foot long, and using DC550 silicone containing 10% w/w of stearic acid as the liquid phase; nitrogen flowing at 18.2 ml min^{-1} was the mobile phase. James' and Martin's recording burette gave the lower, *integral*, chromatogram, and the upper, *differential*, trace was plotted by hand. All modern detectors give a differential trace, in which the area under the peak is proportional to the amount of the component eluted.

After this impetus, gas chromatography advanced rapidly. Ray chromatographed a variety of hydrocarbons, ethers, esters, and alcohols, using a thermal conductivity detector. In 1954 Bradford, Harvey, and Chalkey published their work on hydrocarbons,[7] and this was followed by the work of Keulemans, Kwantes, and Zaul.[8] The technique spread rapidly, and soon commercial instruments became available. Today, the gas chromatograph is a work-horse in thousands of laboratories, and interest is centred more on the results obtainable from the technique than on the technique itself, which can be considered as being well developed.

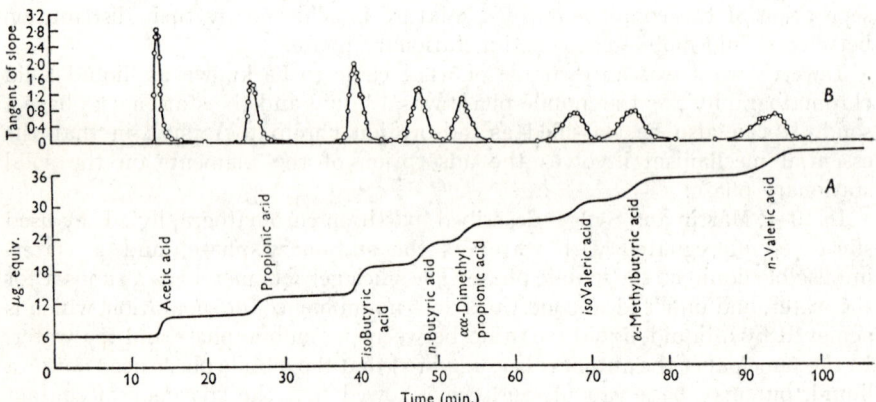

Fig. 1.1   The separation of acetic, propionic, n-butyric, and isobutyric acids and the isomers of valeric acid, showing the complete resolution of all bands and the change in shape of bands as the series is ascended.   Curve A, experimental results; curve B, differential of experimental curve.   Column length is 11 ft; liquid phase is DC550 silicone containing 10% w/w of stearic acid; temperature, 137 °C; rate of flow of nitrogen, 18.2 ml min⁻¹; pressure of nitrogen, 74 cmHg.
(Reproduced from *Analyst*, 1952, **77**, p. 922)

Principles of the Technique

Figure 1.2 represents the basic gas chromatograph. It consists of a supply of carrier gas (A) [nitrogen, helium, or argon can be used, depending on circumstances, which we will consider later]. The carrier gas passes through a fine-control valve (B) to the injection port (C). For liquid samples the injection port consists of a septum of silicone rubber through which the needle of a syringe may be pushed to introduce the sample. The injection area is usually heated to a somewhat higher temperature than the column, so as to ensure flash volatilisation of the sample before it is carried onto the column. If the volatilisation is not virtually instantaneous, the sample is carried onto the column over a period of time, leading to broadening of the peaks and hence a less efficient and unreproducible separation.

If the sample is heat-sensitive, *e.g.* a steroid, then the technique of 'on column' injection is often used, in which the sample is injected directly into the packing at the top of the column. The sample is kept small (a few micrograms dissolved in one or two microlitres of a volatile solvent), to avoid volatilisation problems, and a highly sensitive ionisation detector is used.

To avoid decomposition of the sample, catalysed by the heated metal of the injection port, it is good practice to have a glass or silica lining, which can be removed when it has become fouled by involatile residues.

The column (D) is usually between 1 and 5 metres long and of internal diameter 2 or 4 mm. It may be constructed of glass, stainless steel, or sometimes

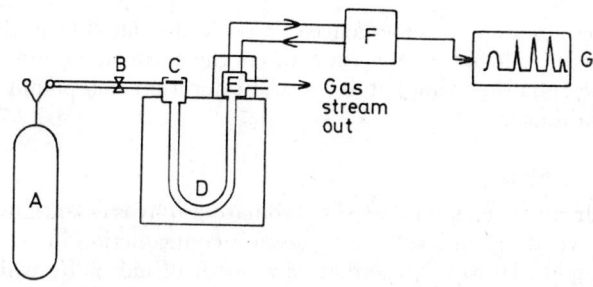

Fig. 1.2.   Components of basic gas chromatograph. A, source of carrier gas; B, fine-control valve; C, injection port; D, column, in oven; E, detector, in oven; F, electronics package controlling (and receiving signal from) the detector; G, potentiometric recorder.

of copper. From the point of view of gas chromatography, glass is probably the best material, as it avoids problems of reactions on metal surfaces which one *can* get with stainless steel and often *does* get with copper. Glass is easier to clean, and can be seen to be clean, and when one is packing the column it can be seen that the packing is complete, with no gaps in the adsorbent. On the other hand, glass is readily broken by the ham-fisted, and is not as easy to plumb into the instrument as a metal column. For this reason, metal columns are widely used for separating relatively unreactive compounds of low polarity.

The column packing is usually a high-boiling organic liquid or grease, coated onto kieselguhr. This, however, will subsequently be discussed in greater detail. The column is contained in an oven, the temperature of which can be closely controlled.

At the column exit is the detector (E), usually enclosed in its own detector oven. The earliest widely used type was the thermal conductivity detector, which detects the emergence of a peak by the change in thermal conductivity of the gas stream flowing from the column. Later followed the much more sensitive ionisation detectors (see section on detectors). The detector is controlled by an electronics package (F) and the signal fed in a suitable form to the potentiometric recorder (G). A 10 millivolt recorder is usual for most instruments. The chart moves at a constant speed and the pen draws a differential chromatogram. This means that, when no component is emerging, a straight line is drawn along the datum line of the chart. As a component emerges, the pen is deflected by an amount proportional to the rate at which the component is emerging, producing (theoretically) a Gaussian peak as the amount of the component increases gradually to a maximum and then falls back to zero. The area under the peak is proportional to the amount of material eluted. One component gives rise to only one peak, but if two or more components have not been separated by the column they will all be present in the same peak. Hence there are at least as many components in the sample as there are peaks in the chromatogram, but there may be more.

The time required for the emergence of a peak, measured from the time when the sample is introduced to the time when the peak maximum is drawn, is known as the retention time; it is constant for that compound under those operating conditions.

Involatile Samples

Since it has already been seen that gas chromatography is essentially a technique for analysing volatile samples, it may seem a contradiction in terms to talk of involatile samples. However, there are a number of means by which involatile materials may be converted into a volatile form, in which they may, with advantage, be subjected to gas chromatography.

The main methods available for this conversion are pyrolysis, by which the material is broken into a number of volatile fragments when heated, and chemical ones. There are two groups of chemical methods; derivative formation, whereby a compound is converted (usually) into a single volatile derivative, and chemical degradation, where a process such as hydrolysis is used to break down a macromolecular substance into smaller volatile materials.

We will consider the chemical methods first, and, since they do not form the main subject of this monograph, the treatment will be brief. The interested reader is referred to the literature for greater detail.

Formation of Derivatives

This technique is particularly valuable in the gas chromatography of carbohydrates. These are of low volatility because of the hydrogen-bonding effects between the large number of hydroxy-groups they contain. However, they can be made more volatile by masking these OH groups, by converting them into esters or ethers which are sufficiently volatile and thermally stable to permit analysis by gas chromatography.[9] Popular derivatives are O-trimethylsilyl ethers,[10] O-acetyl esters,[11] and O-trifluoroacetyl esters.[12]

Chemical Degradation

This is also useful for certain materials. Dried paint and oil films have been analysed by Mills,[13] who hydrolysed the triglycerides with alcoholic KOH, extracted the fatty acids liberated, and then converted them into the methyl esters with diazomethane prior to analysis on an Apiezon L column. He found that different oils could be characterised by the palmitate to stearate ratio, even when they had been dried for many years. The ester linkages in alkyd paints can also be broken down by aminolysis.[14]

In the limited cases where the above techniques can be applied, they are of great value. The reactions taking place can be much more carefully controlled than can pyrolysis, and the fragments are, on the whole, more readily related to the original material than are the products of pyrolysis. On the other hand, the majority of polymeric or macromolecular substances are not amenable to chemical treatment, and if one has a sample of a milligram or less of completely unknown polymer then pyrolysis–gas chromatography can be considered the method of choice for identifying it.

Pyrolysis–Gas Chromatography (p.g.c.)
Pyrolysis is the conversion of a compound into one or more other compounds solely by the use of heat. When combined with gas chromatography it is a powerful tool for the analysis of organic materials, particularly of small samples of polymeric material that are otherwise intractable. This technique will be fully discussed in later chapters.

Detectors

The qualities of the ideal detector are well agreed, as is the fact that no practical detector possesses them all. Because of this, it is necessary to select the detector with those attributes most important for the work to be done.

The ideal detector is inexpensive, robust, easily made (or readily available commercially), and has a rapid response to all materials, the response being linear over the whole chromatographically useful concentration range. The detector should have a rapid recovery from overloading. The response should be highly sensitive to trace components, reproducible, and preferably related in a simple manner to either the weight or number of moles of material passing through it, this relationship being the same for all compounds. The detector should have a high signal to noise ratio. The sensitivity of a detector, S, was defined by Young[15] as the ratio of the change in response R to the change in the quantity measured, Q:

$$S = \Delta R / \Delta Q \tag{1}$$

This can be written as:

$$S = PF/M \tag{2}$$

where P = peak area/mV min
F = flow rate of carrier gas in detector/ml min^{-1}
M = number of millimoles of component

Hence S is expressed in units of ml mV millimole^{-1}.
The limit of detection, Q_0, is given by:

$$Q_0 = 2R_n/S \tag{3}$$

on the assumption that the peak has to be twice the noise level R_n to be observable. R_n is measured in millivolts.

Dimbat, Porter, and Stross[16] defined sensitivity with a similar formula:

$$S = A \ C_1 \ C_2 \ C_3 / W \qquad (4)$$

where S = sensitivity/ml mV mg^{-1}
A = peak area/cm^2
C_1 = recorder sensitivity/mV (cm of chart)$^{-1}$
C_2 = chart rate/min cm^{-1}
C_3 = flow rate of carrier gas at exit/ml min^{-1}
W = weight of sample/mg

This formula differs from that of Young only in its use of weight instead of number of moles of sample. As such, it is probably more relevant to detectors whose response is a function of sample weight rather than of the number of moles of sample.

The Katharometer Detector
The first widely used detector was the thermal-conductivity cell or 'katharometer'. In its simplest form the katharometer consists of a platinum wire, heated by the passage of an electric current, positioned in the carrier-gas stream. The temperature, and hence the resistance of the wire, depends on the current in the wire, the rate of gas flow, and the thermal conductivity of the gas. The wire forms one arm of a Wheatstone bridge, which is balanced so that the point of balance corresponds to the zero line on the recorder. When a component is eluted, the thermal conductivity of the gas changes, causing the temperature and hence the resistance of the wire to change. The amount of unbalance of the bridge is registered by the recorder. The choice of carrier gas is important. Nitrogen has a thermal conductivity close to those of many organic compounds, so that distorted, split, or negative peaks may be obtained, and the sensitivity is low. However, helium and hydrogen have thermal conductivities considerably higher than those of other gases or vapours, so that a much greater response is obtained when using one of these as carrier gas. Helium is the more popular because of its nonreactivity and because of the possibility of an explosion with hydrogen. However, helium is expensive, the detector is less sensitive than with hydrogen, and when flow rates of, say, 50 ml min^{-1} or less are being used, the possibility of explosion when hydrogen is the carrier is remote. The great advantages of the katharometer are its simplicity, the fact that it responds to all materials (making it useful, in particular, for the analysis of permanent gases and water), its stability, and the fact that it does not destroy the sample. Its linear range is limited, but adequate for most applications. It is susceptible to any slight change in the flow rate or in the temperature of the carrier gas or surroundings. For this last reason it has to be kept in a thermostatted oven. The biggest drawback of

the katharometer detector, however, is its low sensitivity. For this reason it has been largely replaced for high-sensitivity work by the ionisation detectors described below.

A basic characteristic of the katharometer is that its response is proportional to the concentration of the eluted component in the cell rather than to the absolute amount. For this reason the sensitivity can be increased by reducing the cell volume. Some modern detectors have volumes as low as a few microlitres, but the usual volume is around 0.5 ml.

The response of the katharometer increases rapidly as the current in the wire is increased. However, this causes increased noise, and is at the expense of filament life. There is also a risk of decomposing the eluted component on the hot wire, though this can be reduced by coating the filament with PTFE.

Katharometer design has been the subject of a large number of papers, e.g. by Brooks et al.[17] and Keppler et al.,[18] Ryce and Bryce,[19,20] Keulemans, Kwantes, and Rijnders,[21] and Schmauch.[22] A more recent review is that by Lawson and Miller,[23] and a useful general review of detectors is that by Gough and Walker.[24]

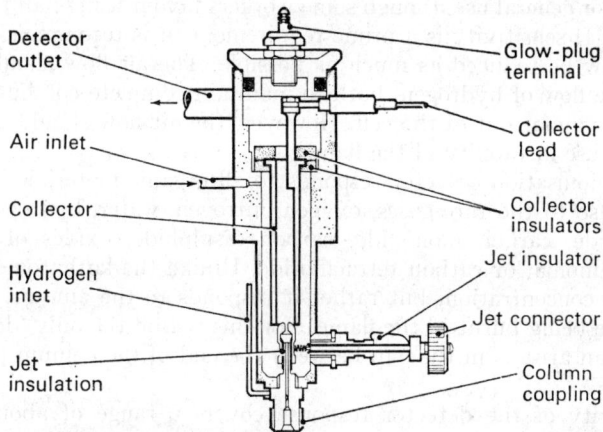

Fig. 1.3. A flame ionisation detector (F.I.D.)
(Reproduced, with permission, from 'Introduction to Gas Chromatography', by R. K. Andrews, Pye Unicam Ltd., Cambridge, 1970).

The Flame Ionisation Detector (F.I.D.)
The flame ionisation detector (F.I.D.) is probably the most popular modern detector, and is rapidly replacing the katharometer for use in the chromatography of organic samples. The column effluent is mixed with hydrogen and the mixed gases are burnt at an electrically insulated jet (see Figure 1.3). Air is supplied to the detector for the combustion. Positioned over the flame is a

cylindrical metal collector electrode, which is also insulated from the body of
the detector. A potential difference is applied between the collector electrode
and the jet. The thermally induced ions in the flame allow a current to flow
between the electrodes. It is usual for the jet to be negative, as the transport
of electrons across the gap is easier than the transport of larger positive ions.
The current is fed to an electrometer amplifier and thence to a recorder. The
small current flowing when only hydrogen is burning at the jet is called the
'standing current'. When an organic substance is eluted from the column and
is burnt in the flame there is a large increase in the number of ions in the flame,
and hence an increase in the current flow, which is monitored by the recorder.
The response for a given amount of material increases with the polarising
voltage at low voltages, but as the voltage is increased the cell reaches a
saturation level, above which the current is independent of the applied
voltage.[25] The saturation voltage depends on the dimensions of the cell, but,
in general, 200 V will saturate the cell regardless of its dimensions or the
operating conditions.

The sensitivity also depends on the flow rate of hydrogen. The lower the
hydrogen flow the better the sensitivity. A 1:1 ratio of hydrogen to carrier
gas is usual for general use, though some workers favour using half that amount
of hydrogen. If sensitivity is a prime requirement, it is recommended that the
hydrogen flow be reduced as much as possible. The air flow should be about
ten times the flow of hydrogen, both to maintain complete combustion and to
sweep water vapour out of the cell. However, the air flow should not be great
enough to cause instability of the flame.

The flame ionisation detector responds to all organic materials. It has little
or no response to the rare gases, oxygen, nitrogen, water, hydrogen sulphide,
carbon dioxide, carbon monoxide, carbon disulphide, oxides of sulphur or
nitrogen, ammonia, or carbon tetrachloride. Unlike the katharometer, it does
not measure concentration, but rather it responds to the amount of material
per unit time being burnt in the flame. For this reason the only 'dead volume'
effect that can arise is in the gap between the end of the column packing and
the jet.

The linearity of the detector response covers a range of about 10^7. This
means that the detector can be calibrated with amounts of material that are
relatively easy to measure and the factor can be reasonably relied on for
amounts at trace levels. The response can be calculated, at least for simple
hydrocarbons, on the basis of the carbon content of the material. The response
correction factor C is given by:

$$C = M_x / 12 C_x \qquad (5)$$

where M_x = molecular weight of component x

C_x = number of carbon atoms in component x

Ettre[26] showed that the equation holds for paraffins, olefins, cycloparaffins,
acetylenes, and aromatic hydrocarbons. The equation is less satisfactory for
oxygenated compounds.

The flame ionisation detector is highly sensitive. Ongkiehong[25] quotes a practical detection limit of 0.001 p.p.m. v/v n-butane in the effluent stream. In fact the detection limit is determined in practice more by the vapour pressure of the stationary phase than by the detector itself. For this reason there is probably little point in looking for a more sensitive general detector. The quest for greater sensitivity must be in the direction of specific detectors which respond to a particular atom or group in the eluate molecule and are unaffected by column bleed.

The Electron-capture Detector (E.C.D.)

The electron-capture detector is a very valuable specific detector of very high sensitivity. It was first introduced by Lovelock and Lipsky[27] as a qualitative detector for distinguishing between such molecular species as ketones, alcohols, esters, ethers, and halides. However, this application is now not nearly as important as is its extreme sensitivity to halogen-containing substances, which gives it great utility in the analysis of pesticide residues.

The detector (Figure 1.4) consists of a source of ionising radiation. This is usually either a tritium-containing foil or one containing ^{63}Ni, formed into a cylinder surrounding a probe electrode. Under the influence of the radiation, free electrons and positive ions are generated in the carrier gas (usually nitrogen). Under the influence of a potential difference between the source and the probe electrode, a current flows in the cell. Because of the high mobility of the free electrons compared with that of the positive ions, virtually no recombination occurs. If a compound which can capture electrons is introduced into the cell, the electrons combine with these molecules to form negative ions, which then combine with the positive ions, causing the current in the cell to fall. The detector response follows a relationship of the Beer's Law type:

$$I = I_0 \exp\left(-ack\right) \tag{6}$$

where I = current when electron-capturing material is present
I_0 = current when no electron-capturing material is present
a = electron-capture cross-section of material
c = concentration of material
k = proportionality constant related to the geometry of the cell and the operating conditions.

The detector can be operated in two ways: in the d.c. mode, in which a constant voltage is applied between the electrodes; or in the pulsed mode, in which the voltage is applied as a pulse of around $1\,\mu s$ duration with a space of between 5 and $150\,\mu s$ between the pulses. In the pulsed mode the sensitivity of the detector is better and the linear range greater than in the d.c. mode.

The linear range is of the order of 10^3, which is much less than that of the flame ionisation detector or of the katharometer. It can be a difficult detector to use, as it is readily overloaded, taking days to recover from a serious overload of a chlorinated material, which may cause permanent damage. It is often used in parallel with a flame ionisation detector, the gas stream being split into two parts, one going to each detector. The F.I.D. will detect all the organic compounds in the eluate, while the electron-capture detector will pick up only the electron-capturing materials, readily detecting nanogram or even picogram amounts of pesticides such as gamma-BHC, aldrin, dieldrin, endrin, and DDT.

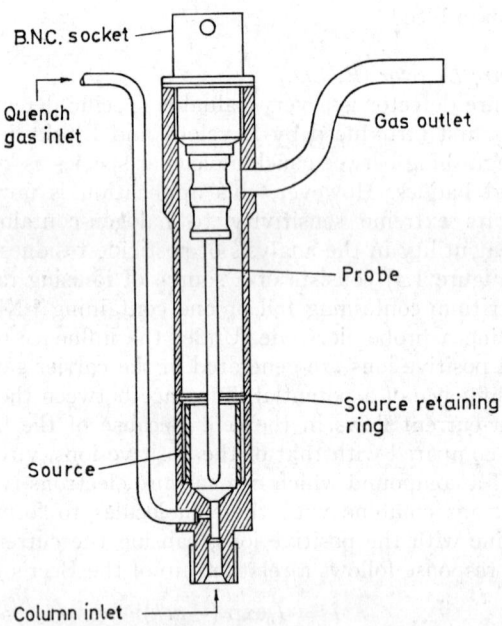

Fig. 1.4. An electron-capture detector (E.C.D.) (Reproduced, with permission of Pye Unicam Ltd., from *Column*, 1966, **1**, No. 3)

The Phosphorus Detector

This detector is an adaptation of the flame ionisation detector. If a sodium salt is placed above the flame of the flame ionisation detector, or if the jet tip is made of compressed salt, the salt becomes thermally ionised, producing a constant current between the electrodes some hundred times greater than the usual standing current. If a phosphorus-containing compound is introduced into the flame, the ionisation efficiency is greatly increased, causing a rise in the cell current. Parts per billion of phosphorus compounds can be detected in $1 \mu l$ samples, and under appropriate operating conditions the detector is specific to phosphorus. However, with appropriate modifications it will also detect halogen- and nitrogen-containing compounds with greater sensitivity than the flame ionisation detector.[28]

The Mass Spectrometer as a Detector

From the point of view of the amount of data it gives, the mass spectrometer can be considered the ultimate detector for use in gas chromatography. It can produce a conventional chromatogram, detect all volatile materials, and simultaneously produce a mass spectrum for each peak, from which the material comprising the peak may be identified. In addition, the combined gas chromatograph–mass spectrometer (G.C.–M.S.) has a sensitivity approaching that of a flame ionisation detector. However, there are drawbacks. A mass spectrometer is expensive, and it is not unusual for one to cost between 10 and 20 times as much as a gas chromatograph. Smaller, cheaper instruments are available, but with a lower performance. Because of the complexity of the technique, the instruments have tended to be under the control of mass spectroscopists rather than gas chromatographers. The former tend to regard a gas chromatograph as a specialised inlet system to the mass spectrometer, rather than the mass spectrometer as a detector for the gas chromatograph. Now that mass spectrometers have appeared designed specifically for use in gas chromatography, and combined G.C.–M.S. units are available commercially, problems in gas chromatography are being looked at more and more, and many elegant solutions to hitherto daunting problems have emerged.

The technique will be considered in greater detail in Chapter 4.

There are a considerable number of other detectors described in the literature. Some are of historical interest only, or are of limited application. The detectors described will cover most uses for gas chromatography. The flame ionisation detector probably covers 90% of them, including the great majority of applications of p.g.c.

Columns

The column is the heart of the gas chromatograph, and a good column produces, on the whole, good results. It may seem paradoxical that, in a sophisticated instrumental technique, the essential part of it is usually hand-made by the analyst himself. The unwary should avoid being taken in by gas chromatograph manufacturers' literature accompanied by impressive chromatograms 'as produced by our instrument'. The column achieved the separation, while the instrumentation is essentially peripheral.

Physically, the column is a long tube holding the stationary phase in such a manner as to present as large a surface area as possible to the mobile phase, so as to allow as rapid an attainment of equilibrium as possible.

In gas–liquid chromatography, the stationary phase is a relatively non-volatile liquid spread in a thin layer over a granular adsorbent of high surface area, which is then packed evenly into the column. In capillary or open tubular columns the stationary phase is spread on the inside surface of the column itself, without the use of a granular solid support.

In gas–solid chromatography, the separation mechanism involves adsorption and desorption at the surface of a solid stationary phase of high specific surface area. Such materials include charcoal, silica gel, molecular sieves, and the more recent organic polymers such as Porapak (Waters Associates) and Chromosorb (101—107) (Johns-Manville).

Occasionally, a combination of gas–liquid and gas–solid chromatography is of value. This usually involves the modification of a solid stationary phase with a low loading of a liquid phase. However, gas–liquid chromatography is the most commonly used form, mainly because, with the enormous number of liquid stationary phases available, it is easier to tailor-make a column to suit a particular application.

Retention Times

The retention time of a solute on a column is a fundamental constant related to the partition coefficient of the solute in the stationary phase. Although it is usually the retention time that is referred to, the retention volume, *i.e.* the volume of carrier gas required to elute the solute under consideration, is a more fundamental parameter. The retention volume, V_R, is given by equations (7) and (8).

$$V_R = Vt_R \tag{7}$$

where V = volume flow rate
t_R = retention time

$$V_R = L\,(a + m\beta) \tag{8}$$

where L = length of column
a = volume of gas per unit length
m = weight of stationary phase per unit length
β = partition coefficient
= wt. of solute per g of stationary phase/wt. of solute per cc of gas phase.

Equation (8) can be rewritten as:

$$V_R = aL + W\beta \tag{9}$$

where W is total weight of stationary phase and aL is the 'dead volume' of the column. The value of 'aL' will be unique to the particular column, as it depends on the way in which the column has been packed. It can be seen that V_R (and hence the retention time) depends only on the column parameters and the partition coefficient, which in turn is temperature-dependent.

Because of the pressure drop along any practical column, the flow rate increases along it. Because of this, the relation (7) is an oversimplification, and a correction factor has to be applied:

$$V_R \text{ (corrected)} = jV_R \text{ (measured)} \tag{10}$$

where j is known as the gas compressibility factor and is given by equation (11). This equation assumes that the flow rate is measured at the column outlet.

$$j = 3[(P_1/P_0)^2-1]/2[(P_1/P_0)^3-1] \qquad (11)$$

where P_1 = inlet pressure

P_0 = outlet pressure

A useful quantity is the 'Specific Retention Volume', V_g, which is corrected for the gas volume of the column, a measure of which is given by the time (t_0) taken for an inert gas to travel through the column:

$$V_g = jV (t_R - t_0)/W \qquad (12)$$

Because of the accurate measurements of column variables as the subsequent calculations involved in obtaining such quantities as V_g, many people prefer to quote relative retention times. It has been recommended by IUPAC that relative retention data should use one or more of eight standard compounds for comparison purposes. These are:

n-butane	naphthalene
2,2,4-trimethylpentane	methyl ethyl ketone
benzene	cyclohexanone
p-xylene	cyclohexanol.

The relative retention usually quoted is made independent of column variables by chromatographing the material and the standard under the same conditions, *i.e.* chromatograph a mixture of them both, and correct for t_0, the time taken for an unadsorbed material such as air to traverse the column.

$$i.e. \text{ relative retention} = (t_R^x - t_0)/(t_R^{std} - t_0) \qquad (13)$$

With a katharometer detector, t_0 is measured from the 'air peak' which always starts the chromatogram. A flame ionisation detector does not give an air peak. However, t_0 can be calculated by extrapolation of a plot of log t_R vs. carbon number for a homologous series of lower n-paraffins. Peterson and Hirsch[29] give the following procedure.

Chromatograph three hydrocarbons whose carbon numbers (n) satisfy the relation:

$$n_2 - n_1 = n_3 - n_2 \qquad (14)$$

then the distance y from an arbitrary point to the air peak is given by:

$$y = (x_2^2 - x_3 x_1)/(x_3 + x_1 - 2x_2) \qquad (15)$$

where $x_{1,2,3}$ is the distance from peaks 1,2,3, respectively, to the point y. Since very many pyrograms include those of a series of n-paraffins among the peaks, this can be a convenient measurement to make. (See section on G.C. standardisation in Chapter 5).

Column Efficiency

In the ideal column there would be no broadening of the solute band as it travelled through the column. However, in a real column the band is always broader as it emerges than it was when it entered the column. The main broadening factors are the finite time taken for equilibration of the solute between the two phases, longitudinal diffusion of the solute in the column, and the geometry of the packing, *i.e.* the solute molecules can take different routes, of different lengths, through the packing. The more efficient a column, the more nearly it approaches an ideal one. Theoretically, the shape of the emerging solute peak is a Gaussian curve, since there is a normal distribution of solute molecules around the mean, *i.e.* the peak maximum. This normal distribution is measured by the standard deviation σ. In this instance σ is equal to half the width of the peak measured at the points of inflexion, or alternatively, width at half height $= 2.354\sigma$. The basic measure of a normal distribution, mathematically, is the variance, σ^2. The total variance can be considered as the sum of the variances contributed by each element of column length. These will be the same in a uniform column. The total variance is therefore directly proportional to the column length, L. The constant of proportionality, h, is given by:

$$h = \sigma^2/L \tag{16}$$

and the constant h is called the 'height equivalent to a theoretical plate' (HETP). The concept of plate height is taken from the theory of distillation columns, and although it has no physical reality in gas chromatography it provides a useful numerical description of the efficiency of a column.

The number of plates in a column, N, is given by:

$$N = l^2/\sigma^2 \tag{17a}$$

$$\text{or } N = 5.54 \ l^2/w^2 \tag{17b}$$

where $w =$ width of the peak at half height
       $l =$ distance along chart, in the same units
The total contribution to peak broadening of all factors (i) is given by:

$$\sigma_{\text{total}}^2 = \sum_i \sigma_i^2 \tag{18}$$

The actual effect on peak width is governed by σ, the standard deviation. The total broadening therefore is given by:

$$\sigma = (\sum_i \sigma_i^2)^{\frac{1}{2}} < \Sigma_i \ \sigma_i \tag{19}$$

Because of this, the broadening is dominated by the largest factor, and minor factors have a negligible effect; for example, take two factors having individual standard deviations of 10 and 2 seconds. The total standard deviation is:

$$\sqrt{(10^2 + 2^2)} = \sqrt{104} = 10.2\,\text{s} \tag{20}$$

The minor factor is 20% of the major factor, but its final contribution to the peak broadening is only 2%.

The broadening factors and their effect on the plate height h have been treated mathematically[30] to give the expression for h usually called the van Deemter equation:

$$h = 2\lambda\, d_p + (2\gamma\, D_g/u) + [8d_f^2 k/\pi^2 D_1\, (1 + k)^2]u \qquad (21)$$

where λ = a constant that is related to the packing
d_p = average grain diameter of the packing
γ = tortuosity factor
D_g = diffusion coefficient of solute in the gas phase
D_1 = diffusion coefficient of solute in the stationary phase
d_f = thickness of liquid film
u = linear gas velocity
k = partition ratio

This can be written as:

$$h = A + (B/u) + Cu \qquad (22)$$

where A = $2\lambda d_p$
B = $2\gamma D_g$
C = $(8/\pi^2)[k/(1 + k)^2](d_f^2/D_1)$

A, B and C are constants for a particular column and are not functions of the flow rate. The equation gives a hyperbola when h is plotted against u, which is obtained most readily from the retention time (t_0) of an unretained solute, *i.e.* from the 'air peak'. Then,

$$u = t_0/L \qquad (23)$$

where L is the column length

The van Deemter equation is a minimum when:

$$B/u = Cu \qquad (24)$$

$$\text{Hence } u_{\text{minimum}} = (B/C)^{\frac{1}{2}} \qquad (25)$$

$$\text{and } h_{\text{minimum}} = A + (2BC)^{\frac{1}{2}} \qquad (26)$$

The conditions for efficient chromatography can be readily ascertained from the van Deemter equation. Obviously, one should try to keep all three terms as low as possible. The first term is independent of the flow rate, while the second and third terms are dependent on flow rate, being respectively inversely and directly proportional; this makes the plate height very much dependent on the flow rate, which should be adjusted to the optimum value. However, in practice, the minimum of the hyperbola is very shallow, particularly on the side of increasing values of u, so that it is often possible to double the optimum flow rate and still only marginally increase the plate height. The equation shows h rising much more rapidly as u is decreased below the optimum value, and operating on this part of the curve should be avoided both because of the inefficiency of the column and because of the long analysis times inherent in a low flow rate. The term A was described by van Deemter and his co-authors as the 'eddy diffusion term', and is governed by the flow pattern of the carrier gas through the irregular channels in the packed column. The term λ is related to the structure of the packing, and, if the particles are regular in size and evenly packed, its value approaches 1. Since the grain diameter d_p is small, the term A is also small, and its presence in practical h vs. u plots probably indicates dead space in the system other than in the column.

The term $(B/u)=(2\gamma D_g/u)$ is governed by longitudinal diffusion of solute molecules in the gas phase. Since the diffusion is greater the longer the solute remains in the column, it is inversely proportional to the flow rate. The tortuosity factor (γ) is an experimental quantity whose value is usually between 0.5 and 1, and governed by the twists and turns of the solute molecules in the channels through the packing. Since practical columns are almost always operated at flow rates above optimum, the term B has a negligible effect on the plate height. This makes C the most important term. This term includes the resistance to mass transfer in the liquid phase:

$$C = (8/\pi^2) \, [k/(1 + k)^2] \, (d_f^2/D_l) \tag{27}$$

where $k = \beta V_L/V_G$ and V_L and V_G are volumes of liquid and gas phases, respectively, and β is the partition coefficient. If β is increased, $k/(1 + k)^2$ is decreased; hence a column is more efficient for materials with long retention times than it is for materials with short ones.

The liquid loading has a contradictory effect. An increase in the liquid loading increases k, and hence diminishes $k/(1 + k)^2$. However, the square of the liquid thickness appears in the numerator. In practice, the liquid thickness factor would appear to be more important, since it is a fact that columns with a low loading of stationary phase are more efficient, other factors being equal, than those with a high loading.

The van Deemter equation is very valuable for the qualitative picture it gives of the interaction of the various factors in gas chromatography. However, it does not stand up very well to detailed quantitative examination, due to the many simplifications contained in its derivation. A number of people have examined the van Deemter equation and there have been a number of suggested improvements. However, these are at the expense of simplicity, and to the average practising gas chromatographer they may seem to obscure rather than to illuminate the overall picture given by the van Deemter equation of the process taking place inside the column. There are many other papers describing attempts to improve the equation.[31–38]

Capillary Columns

Probably the most significant advance to come out of these theoretical considerations of column performance was the invention in 1957 by Golay of capillary or open tubular columns.[39,40] He considered how to reduce zone spreading due to the packing and due to the pressure drop along the column, and came to the conclusion that an unpacked capillary tube with its internal wall coated with a film of stationary phase would be a highly efficient column.

Golay deduced an expression for the plate height (h) of a capillary column, assuming a liquid layer of uniform thickness and laminar gas flow in the column:

$$h = (2D_g/u) + [2k\ d_f^2/3(1 + k)^2 D_1]u + \\ [(1 + 6k + 11k^2)r^2/24(1 + k)^2 D_g]u \qquad (28)$$

where r = radius of tube, and other terms are the same as before. This reduces to:

$$h = (B/u) + C_1 u + C_g u \qquad (29)$$

The term A in the van Deemter equation is absent in Golay's equation. As this term was connected with the packing, this is of course to be expected. $C_1 u$ is governed by the resistance to mass transfer in the liquid phase and is equivalent to the last term in van Deemter's equation. The value of $C_g u$ is governed by the resistance to mass transfer in the gas phase. The absence of this term in van Deemter's equation is considered to be a weakness, and the term is included in modifications of it.[41] As in the van Deemter equation, the term $C_1 u$ is the most important.

It is possible to make capillary columns with far higher plate numbers than is possible with packed columns. Because of the low pressure drop per unit length, capillary columns of great length can usefully be used. Desty has described a 900 foot column of around a million plates.[42]

Because of the very low loading of liquid phase in a capillary column, it can only take a very small sample before being overloaded. This is usually achieved by having a stream splitter, which allows only a fraction of the sample to pass onto the column. Also, a detector with a very small dead volume has to be used. In practice, the flame ionisation detector is excellent, as the only dead volume that can arise in this detector is in the piping between the end of the column and the jet. This can be made very small by suitable design.

Resolution

Resolution (R) is defined by equation (30), where t_2 and t_1 are the retention times of the two components being separated, and $t_2 > t_1$; W_2 and W_1 are the widths measured on the base-line between the tangents at the points of inflexion of the peak (see Figure 1.5).

$$R = 2(t_2 - t_1)/(W_2 + W_1) \qquad (30)$$

For neighbouring peaks it is approximately true that:

$$W_1 = W_2$$
$$\therefore R = (t_2 - t_1)/W_2 \qquad (31)$$

In a Gaussian curve the peaks are 98% separated when $t_2 - t_1 = W_2$, *i.e.* the maxima are separated by a distance equal to the base width; *i.e.* $R = 1$. The peaks are 99.7% separated when $R = 1.5$. In theory, two Gaussian peaks can never be completely separated, as the baseline is an asymptote to the peak.

In practice, however, $R = 1$ is usually considered as separation. The resolution for closely spaced peaks, according to Purnell,[43] is given by:

$$R = \tfrac{1}{4}\,[(\alpha - 1)/\alpha]\,[k/(1 + k)]\,(L/h)^{\frac{1}{2}} \tag{32}$$

where α = relative retention
k = partition ratio for the slower peak
L = column length
h = height equivalent to a theoretical plate

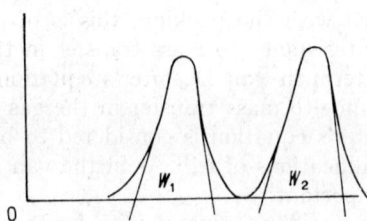

Fig. 1.5. Method of measuring peak width

This equation shows that the resolution is proportional to the square-root of the column length rather than to the length itself. Although the distance separating the peak maxima is proportional to L, the peak width is proportional to $\sqrt{L}$.

The relative retention α is governed by the selectivity of the stationary phase for the two solutes in question. The value of k, the ratio of the amounts of solute in each phase, is a measure of the retentivity of the column, whereas h, the plate height, is governed by the factors discussed in connection with the van Deemter equation. To determine the length of column required for a given separation, substitute $R = 1$ in equation (32).

Then $$L = 16h\,\{[\alpha/(\alpha - 1)]\,[(k + 1)/k]\}^{2} \tag{33}$$

Conclusions:
From the foregoing discussions we now have a good idea of the factors which go to make up an efficient column. It may be of value to summarise these.

(1) Packing
The grains should be of an even size; in particular, all fines should be sieved out. For a column of 4mm internal diameter, 80—100 mesh is most suitable. For a 2mm column, 100—120 mesh is probably best. The finer the mesh size the better is the efficiency at high carrier flow rates, but the permeability of the column drops, and there is loss of efficiency due to the pressure drop across the column, so a compromise has to be reached.

(2) Liquid Loading

The lower the liquid loading the higher the efficiency, but the sample capacity also falls, and at very low loadings the resolution will suffer. A standard column will have between 5 and 20% weight/weight liquid phase on the solid support, though low-loaded columns (containing around 0.5% stationary phase) are used for special applications. If the loading is low there are also difficulties with the adsorption by the solid support, and due to the finite bleed of liquid phase; therefore the column life will be short.

(3) Column Length

The minimum column length to give the required resolution should be used. Resolution increases as the square-root of the length, but analysis time increases as the length increases. Again, a compromise has to be reached. A length of between one and two metres is usual for a packed column.

(4) Temperature

As the temperature is increased, the retention times decrease, but so does the resolution. Normally, the temperature is set to give the minimum acceptable resolution, and so obtain the maximum possible speed of analysis. The shorter the retention time the better is the shape of the peak, *i.e.* the height to width ratio is higher. A practical temperature limit is set by the nature of the stationary phase. At higher temperatures the bleed increases rapidly, thermal breakdown of the stationary phase may occur, and the column life is reduced.

(5) Carrier Flow-rate

The optimum flow rate is given by the minimum in the h *vs.* u plot, and flow rates below this minimum should not be used. However, the right-hand 'leg' of the hyperbola has a fairly low gradient, and in practice it is possible to operate at much higher flow rates with little loss of efficiency. The flow rates usually used in a column of 4mm internal diameter are between 20 and 60ml min^{-1}. In fact, the same separation can often be achieved faster by using a longer column at high flow rate than by operating a shorter one at its optimum flow rate.

Stationary Phases

The choice of stationary phase is the most important step when embarking on the preparation of a column, and the number of phases to choose from is enormous. Although it is usually possible to decide what class of material will be suitable, the final choice is bound to be made on a trial-and-error basis. A very large number of separations is described in the literature, and guidance from the accumulated experience of others should be accepted.

We will consider liquid phases before solid phases. Phases are classified as being polar or non-polar. A non-polar phase will, on the whole, separate solutes according to their boiling points. Such materials are hydrocarbon greases like Apiezon L and some silicone greases or oils. They are useful for hydrocarbons and high-boiling materials such as steroids, where the retentivity of a polar phase would be undesirable. Most important, they are good general-purpose columns on which it is well worth running an unknown sample initially. Columns loaded with Apiezon L and low-polarity silicones (SE30, E301 *etc.*) are of considerable utility in pyrolysis–gas chromatography, particularly where a general-purpose phase is required because of a wide range of samples or because a wide variety of pyrolysis products is expected.

For separations where selectivity for particular groups of compounds is required, one turns to the more polar phases. The general rule here is that like retains (and is hence selective for) like. For instance, the poly(ethylene glycols) will separate alcohols and compounds of similar structure such as aldehydes and ketones very well. Ester phases such as neopentyl glycol adipate and diethylene glycol succinate will separate fatty acid esters. By trying to match the polarity and compound type of the stationary phase to those of the solute, bearing in mind the maximum usable temperatures of the various stationary phases, either a non-polar stationary phase or a polar one is chosen. Probably most important of all, the literature should be searched to see what others have used, and how successful they were. See also Chapter 5.

The Solid Support
The ideal solid support has a large surface area, is chemically inert, thermally stable, is of uniform mesh size, and has a high mechanical strength. Diatomaceous earth supports are the most widely used. This material originates from fossilised diatoms and consists mainly of amorphous silica. It has an open, porous structure, giving it a high surface area. Heat treatment of the mineral produces the various commercial supports available under the names of Celite, Chromsorb, *etc.* Commercial materials which have been washed with acid and alkali and then sieve-sized are available. They can be obtained already silanised or they can be silanised in the laboratory. This involves treating the supports with dimethyldichlorosilane, to react with the –Si–OH groups on them:

$$\text{–Si–OH} + \text{Me}_2\,\text{SiCl}_2 \rightarrow \text{–Si–O–SiMe}_2\text{Cl} + \text{HCl}$$

This is followed by a wash with methanol to remove the chlorine:

$$\text{–Si–O–SiMe}_2\text{Cl} + \text{MeOH} \rightarrow \text{–Si–O–SiMe}_2\text{OMe} + \text{HCl}$$

Glass columns can be similarly treated before filling. The column packing is usually prepared by taking up the stationary phase in a quantity of suitable solvent, stirring in the support to form a suspension, and evaporating off the solvent on a water bath or in a rotary evaporator. The support will completely absorb the stationary phase to give a dry, free-flowing powder, with which the column may be filled by applying a vacuum at the far end and tapping or vibrating the column until no more material will go in. Each end of the column

should be plugged with silanised glass wool. The column is conditioned over-
night with a low carrier-gas flow at around 30 °C above the maximum required
temperature of use. It is then ready for use.

Although diatomaceous earths are the most usual supports, Teflon and glass
microbeads are also used occasionally. There are not nearly as many solid
stationary phases available as there are liquid phases. The traditional solid
phases are materials such as active carbon or charcoal, alumina, silica gel, and
molecular sieves. They are mostly useful in the separation of low-boiling hydrocarbons and permanent gases. Molecular Sieve 5A, activated at between 300
and 500 °C, is very useful for the separation of oxygen and nitrogen, and for
removing water from samples, when in the form of a short pre-column.

With all but the most volatile samples, peak tailing and unreproducible
retention times occur with the solid phases listed above. This severely restricts
their usefulness. However, useful capillary columns have been made by coating
the inside wall with highly dispersed silica in the form of a colloidal sol. They
have a resolution comparable to those of partition capillary columns.[44] The
use of hydrogenated carbon black has also been reported.[45]

One of the most significant developments in gas–solid chromatography came
with the development of porous polymer beads as stationary phases.[46] These
are cross-linked polyaromatic polymer beads whose porous structure gives
them a large surface area. Commercial types include the Porapaks and Chromosorb 101—107. The most widely used is probably Porapak Q (Waters Associates,
Inc.), which is ethylvinylbenzene crosslinked with divinylbenzene. Porapak Q
has a much greater retentivity for solutes than liquid stationary phases, but
its most significant feature is its much lower relative retention of hydroxylated
compounds as against hydrocarbons. For instance, in a mixture of hydrocarbons, alcohols, and water, peaks corresponding to methane, ethylene, and
ethane will appear, followed by water, propylene, propane, methanol,
ethanol, and butane in that order; all separated on a 1 m column at 100 °C.
The peak shapes of both polar and non-polar materials are excellent. The
main drawback of Porapak Q is its retentivity. It will not elute hydrocarbons
with a molecular weight much over 100 in a reasonable length of time. Its
maximum temperature of operation is around 200 °C, but it has no minimum
operating temperature as it has no freezing point, and it will separate the
components of air at −78 °C. However, because of its great versatility in the
separation of all sorts of polar and non-polar materials, Porapak Q is very
useful in the separation of pyrolysis products, as we shall show in subsequent
chapters.

References for Chapter 1

1   Tswett, M., *Ber. Deut. Botan. Ges.*, 1906, **24**, 316.
2   Martin, A. J. P., and Synge, R. L. M., *Biochem. J.*, 1941, **35**, 1358.
3   James, A. T., and Martin, A. J. P., *Biochem. J.*, 1952, **50**, 679.
4   James, A. T., Martin, A. J. P., and Howard-Smith, G., *Biochem. J.*, 1952, **52**, 238.
5   James, A. T., *Biochem. J.*, 1952, **52**, 915.

6   James, A. T., and Martin, A. J. P., *Analyst*, 1952, **77**, 915.
7   Bradford, B. W., Harvey, D., and Chalkey, P. E., *J. Inst. Petroleum*, 1955, **41**, 80.
8   Keulemans, A. I. M., Kwantes, A., and Zaul, P., *Analyt. Chim. Acta*, 1955, **13**, 357.
9   Slonecker, J. H., in 'Biomedical Applications of Gas Chromatography,' Vol. 2, ed. Szymanski, H. A., Plenum Press, New York, 1968, p. 89.
10 Sweeley, C. C., Bentley, R., Makita, M., and Wells, W. W., *J. Amer. Chem. Soc.*, 1963, **85**, 2497.
11 Wells, W. W., Sweeley, C. C., and Bentley, R., in 'Biomedical Applications of Gas Chromatography,' ed. Szymanski, H. A., Plenum Press, New York, 1963, p. 169.
12 Vilkas, M. Jan, H., Boussac, G., and Bonnard, M. C., *Tetrahedron Letters*, 1966, 1441.
13 Mills, J. S., *Studies in Conservation*, 1966, **11**, 92.
14 Esposito, G. G., and Swann, M. H., *Analyt. Chem.*, 1969, **41**, 1118.
15 Young, I. G., in 'Second International Gas Chromatography Symposium of the Instrumentation Society of America,' Academic Press, New York, 1959, p. 75.
16 Dimbat, M., Porter, P. E., and Stross, F. H., *Analyt. Chem.*, 1956, **28**, 290.
17 Brooks, J. Murray, W., and Williams, A. F., in 'Vapour Phase Chromatography,' ed. Desty, D. H., Butterworths, London, 1956, p. 281.
18 Keppler, J. G., Dijkstra, G., and Schols, J. A., in ref. 17, p. 222.
19 Ryce, S. A., and Bryce, W., *Analyt. Chem.*, 1957, **29**, 1386.
20 Ryce, S. A., and Bryce, W., *Nature*, 1957, **179**, 541.
21 Keulemans, A. I. M., Kwantes, A., and Rijnders, G. W. A., *Analyt. Chim. Acta*, 1957, **16**, 29.
22 Schmauch, L. J., *Analyt. Chem.*, 1959, **31**, 225.
23 Lawson, A. E., and Miller, J. M., *J. Gas Chromatog.*, 1966, **4**, 273.
24 Gough, T. A., and Walker, E. A., *Analyst*, 1970, **95**, 1.
25 Ongkiehong, L., in 'Gas Chromatography 1960,' ed. Scott, R. P. W., Butterworths, London, 1960, p. 7.
26 Ettre, L. S., in 'Gas Chromatography,' ed. Brenner, N., Callen, J. E., and Weiss, M. D., Academic Press, New York, 1962, p. 307.
27 Lovelock, J. E., and Lipsky, S. R., *J. Amer. Chem. Soc.*, 1960, **82**, 431.
28 Ives, N. F., and Guiffrida, L., *J. Assoc. Offic. Analyt. Chemists*, 1967, **50**, 1.
29 Peterson, M. L., and Hirsch, U., *J. Lipid Res.*, 1959, **1**, 132.
30 van Deemter, J. J., Zaiderweg, F. J., and Klinkenberg, A., *Chem. Eng. Sci.*, 1956, **5**, 271.
31 Keulemans, A. I. M., and Kwantes, A., in 'Vapour Phase Chromatography,' ed. Desty, D. H., and Harbourn, C. L. A., Butterworths, London, 1957, p. 15.
32 Littlewood, A. B., in 'Gas Chromatography,' ed. Coates, V. J., Noebels, H. J., and Fagerson, I. S., Academic Press, New York, 1958, p. 23.
33 Cheshire, J. D., and Scott, R. P. W., *J. Inst. Petroleum*, 1958, **44**, 74.
34 Golay, M. J. E., and Purnell, J. H., *Ann. New York Acad. Sci.*, 1959, **72**, 612.
35 Lloyd, R. J., Ayers, B. O., and Karasek, F. W., *Analyt. Chem.*, 1960, **32**, 698.
36 Jones, W. L., *Analyt. Chem.*, 1961, **33**, 829.
37 Kieselbach, R., *Analyt. Chem.*, 1961, **33**, 806.
38 Ford, D., Lloyd, R. J., and Ayers, B. O., *Analyt. Chem.*, 1963, **35**, 426.
39 Golay, M. J. E., in 'Gas Chromatography,' ed. Coates, V. J., Noebels, H. J., and Fagerson, I. S., Academic Press, New York, 1958, p. 1.
40 Golay, M. J. E., in 'Gas Chromatography,' ed. Desty, D. H., Butterworths, London, 1958, p. 36.
41 Glueckauf, E., Golay, M. J. E., and Purnell, J. H., *Ann. New York Acad. Sci.*, 1959, **72**, 612.
42 Desty, D. H., *Adv. Chromatog.*, 1965, 211
43 Purnell, J. H., *J. Chem. Soc.*, 1960, 1268.
44 Schwartz, R. D., Brasseaux, D. J., and Mathews, R. G., *Analyt. Chem.*, 1966, **38**, 303.
45 Lloyd, J. B. F., Hadley, K., and Roberts, B. R. G., *J. Chromatog.*, 1974, **101**, 417.
46 Hollis, O. L., *Analyt. Chem.*, 1966, **38**, 309.

2 Pyrolysis Apparatus

The use of pyrolysis for the analysis of organic materials is an old technique. A common form consists of heating the material on a spatula or in a test-tube until decomposition occurs; physical or chemical tests may then be carried out on the decomposition products.[1-3] One of the most simple tests, the smell of burning, may be characteristic, though due regard to the possible production of toxic products should be taken; for example, carbonyl fluoride is produced by heating polytetrafluoroethylene in air.[4]

A closed pyrolysis system can be used from which the more volatile products can be subsequently extracted and analysed by gas chromatography (g.c.).[5] Separation of the pyrolysis stage from the g.c. in this way is useful in mechanistic studies of thermal degradation under vacuum or special atmospheres,[6-9] but for analytical problems, the direct coupling of pyrolyser and g.c. is more common, and it makes for easier handling and rapid results. Only such directly coupled systems will be considered under our heading of pyrolysis–gas chromatography (p.g.c.). We are mainly concerned with the use of the technique in analysis, but it is also useful in kinetic studies of thermal degradation,[10-12] and the two subjects cannot be completely separated. We shall confine our discussion of the kinetic aspects of the subject to the minimum necessary for the understanding of the operation of analytical instrumentation. Techniques in which catalysis (such as in catalytic hydrogenation[10]) and other surface effects[11] are deliberately induced are also outside the scope of this work, though the dividing line is inevitably vague, since catalytic effects are rarely completely absent.

Before discussing individual types of pyrolyers, it is worth considering some general points. The requirements of a good analytical pyrolyser are the production of degradation products that are; (1) as nearly unique to the sample as possible,[12] (2) reproducible, (3) capable of successful separation and elution in the gas chromatograph (G.C.). With some substances, these requirements conflict. Unique degradation products are often of high molecular weight, and are produced at relatively low pyrolysis temperatures. Low temperatures in turn give low degradation rates, and in the case of pyrolysers through which the carrier gas flows continuously (dynamic mode), acceptably high yields of pyrolysis products may only be produced over long periods, leading to broad, incompletely resolved g.c. peaks (this latter effect may be reduced by programming the column oven from a low temperature, at which a plug of pyrolosis products will build up at the top of the column). Cool spots between pyrolyser and column oven become a danger with higher-molecular-weight, less-volatile products, and such products may be difficult to elute at normal g.c. temperatures. During sample heating and cooling, different degradation mechanisms often operate than occur at the nominal pyrolysis temperature. If these different mechanisms give the same product or ratio of products, then sample heating and

23

cooling rates will not be criticial; unfortunately, this is frequently not so. Consequently, the pyrolyser should be capable of heating the sample very rapidly, so that an insignificant proportion of the sample decomposes before the 'pyrolysis temperature' is reached,[13] or the sample must be heated at a readily reproducible rate, so that products formed during heating account for a fixed proportion of the total pyrolysis mixture. The effect of rate of cooling is best minimised by ensuring complete pyrolysis of the sample before the temperature is allowed to fall. Variations in the final composition can be caused by secondary reactions in the hot gaseous products, where reaction rates are likely to be dependent on the concentration. In a throughflow (dynamic mode) pyrolyser, secondary reaction effects in the hot product gases are kept to a minimum by holding the concentration of the product gases down. This can be done by keeping the sample size low and by sweeping the products away rapidly. Fast and reproducible heating rates can only be achieved in thin samples, since the rate of heating will depend on the rate of conduction if the thickness is significant. It can be seen, therefore, that with any throughflow pyrolyser the reproducibility of product composition will be better with small samples, preferably of the order of a few micrograms or less, depending on the particular sample and pyrolyser.

The design of a pyrolyser for applications where comparison with pyrograms obtained from other units is not necessary requires only that a useful and precise temperature, usually in the region 500–800 °C[12,14] (not necessarily accurately known) is achieved at a fixed or near instantaneous heating rate in an environment constant with respect to gas flow and volume. If the pyrolysate from one pyrolyser is to be compared with that from another, then the heating rates and the final temperatures of the two pyrolysers must be similar to ensure matching of the primary degradation reactions. The dimensions, flow, and nature of carrier gas and temperature gradients within the units must be similar so as to ensure that equivalent amounts of secondary reactions occur, and the amounts of materials which may come into contact with either hot sample or hot pyrolysate must be similar, or else known not to catalyse thermal reactions.

Clearly, great advantages can be achieved if pyrograms obtained in different laboratories can be usefully compared. Part of the reason why this is not more frequently done must lie in the difficulty in fulfilling the above requirements in pyrolyser design. The importance of these design variations can be minimised by the user by ensuring small sample sizes and fast flow rates of the carrier gas, so that only primary degradation products are formed in significant amounts and catalytic effects are negligible.[15] Perhaps the best way to reduce the importance of design variations in heating rate is to make the rate as fast as possible;[16] this in turn means that the final temperature must be constant, as the bulk of pyrolysis will then occur at the upper temperature.[17]

A comprehensive review of the technique of p.g.c. has been written by Levy,[18a] and a more recent short one by Walker.[18b] Other reviews in English, placing more emphasis on applications, have appeared from time to time.[4,19-21] We will, therefore, concentrate largely on the most recent literature, whilst attempting to maintain a balanced view of the subject as a whole.

The types of pyrolysis apparatus used for p.g.c. of organic material can be divided into those designed for pyrolysing solids and involatile liquids and those for gases and volatile liquids. The former are currently of most importance in analysis, and will be considered first.

Apparatus for the Pyrolysis of Solids and Involatile Liquids

There are three main types of pyrolyser for solids and involatile liquids in common use today; *viz.* hot-filament, Curie-point, and furnace;[22] other, less common, types of heater will be discussed towards the end of this chapter.

Pyrolysers are classed as being of pulse-mode or continuous-mode types;[16] hot-filament and Curie-point methods operate in the pulse mode, and furnaces in the continuous mode. There are important differences in the means of application of heat in the two modes which can lead to differences in product composition. A pulse-mode pyrolyser usually operates by heating from a central source, commonly a metal wire with which the sample is in intimate contact, the whole unit being flushed by carrier gas. The heat source is raised quickly from ambient to pyrolysis temperature. As soon as vaporised pyrolysis products escape from the sample they are likely to enter a cooler area and be swept away into even cooler regions; this rapid cooling will minimise the amount of products from secondary reactions. In a continuous-mode pyrolyser (furnace unit) the heat source is the wall of the unit, maintained at the pyrolysis temperature. Since the sample is often supported free of the walls, the vaporised pyrolysis products will leave the relatively cool sample and move into a hotter region; this condition will continue throughout the duration of the pyrolysis unless a strongly exothermic reaction is initiated in the sample. The chances of secondary reactions are therefore increased, and their extent at a given temperature will be more dependent on the residence time (*i.e.* the time spent in the pyrolyser) of the products, which in turn is dependent on the internal dimensions of the unit and the flow rate of the carrier gas. These design aspects have been considered by Levy.[16] We will now describe each type of unit in detail.

Continuous-mode, Furnace or Microreactor Type

Perhaps the most readily available pyrolyser is simply an injection-port heater combined with a sample injector or holder. Without modification, the heater will probably only operate at low, ill-defined temperatures. Burrows and Callam[23] differentiated between several antibiotics with a system of this type held at $380 \pm 10\,°C$, using a solids injector to introduce the sample. Contamination of the walls is always a problem with pyrolysis, particularly if the sample contacts the walls directly. This problem is, however, shared by injection ports used for their normal purpose, so that replaceable liners are often available.

Many analyses require higher temperatures and more precise temperature control; hence the use of purpose-built pyrolysers. Major features of a furnace pyrolyser are:

(i) the tube through which carrier gas flows and in which the decomposition occurs;

(ii)   the heat source, normally a furnace employing direct resistive heating but sometimes consisting of a hot fluid bath in which the tube is immersed (a refluxing liquid gives good temperature control); Romováček and Kubát[24] used a bath of molten tin;

(iii)   a method of sample delivery from a cool area into the hot zone.

The tube should be constructed of non-catalytic material and the geometry should be such as to promote rapid removal of pyrolysis products. The heat source should give a simple and constant temperature profile in the tube. The sample-delivery device should not hinder rapid heat-up of the sample, as we have seen that heating-up time is an important parameter.

The design of the sample-delivery device warrants particular care, and several approaches that have been employed are discussed below.

(a) *Gravity Feed*. The sample may be dropped from a cool stand-by area into the hot zone, for example by a magnetic release system; this will usually require the sample to be held in some kind of container. Liddicoet and Smithson[25] used a stainless-steel cup; tin sample containers melting at the pyrolysis temperature have been used,[24] but the latent heat required is likely to slow the heat-up rate, and a build-up of pyrolysis residues will occur if the apparatus is not frequently cleaned out. Hewitt and Witham[26] used capillary tubes attached to iron slugs, so that they could be withdrawn by means of a magnet after pyrolysis; if the bulk of the pyrolysis residue remains in the sample container, this procedure will minimise the contaminant level, and it is easily automated.[24] Care must be taken in the design to ensure that carrier gas flows over the pyrolysing sample and not merely above it, leaving a 'stagnant' atmosphere in which secondary reactions will predominate. Hewitt and Witham[26] used a vertical furnace, the capillary tubes falling onto the top of a packed section of the furnace, so that the sample was well swept by the carrier gas. The large surface area of a packed furnace allows good thermal contact of heat source and carrier-gas stream but also greatly increases the chances of surface-catalysed reactions, which will vary in extent, at least until a good coating of inert pyrolysis products has built up on the packing. Packed furnaces are therefore of limited use.

(b) *Magnetic Push-rod*. A novel means of introducing several samples in turn into the furnace was described by Ettre and Varaldi,[27] and will serve as an example of the general method. A 'train' of sample boats separated from each other by nickel rods are placed in a side-arm. The first sample is pushed by the following nickel rod, which is moved by an external magnet, into a horizontal tube which passes through a furnace. The tube is cool at the junction with the side-arm and the sample is pushed home into the furnace by means of a plunger, again operated magnetically. After pyrolysis, the sample boat can be removed by the plunger to a storage sump. An adaptation of this apparatus was recently used by Armitage[28] for quantitative polymer analysis.

A specialised furnace pyrolyser for use with organic material locked in an inorganic matrix has been used in the organic analysis of meteoric and lunar matter.[29-31] Difficulties in holding the sample were avoided by moving the furnace to the sample rather than *vice versa* (the usual method). About 0.05 g of sample was used and placed in a quartz sample tube (12 cm long, $\frac{1}{8}$ in o.d., 5/64 in i.d.) which served as the carrier-gas inlet to the G.C. The pulverised sample was held in position 2 cm from the G.C. inlet by a plug of quartz wool. The furnace consisted of a hollow silver rod fitting closely around the sample tube and with a heating coil wrapped around the outside of the rod. An outer quartz sleeve provided thermal insulation. The furnace was heated to equilibrium temperature upstream of the sample and then slid down to surround the sample; it heated it to a range of temperatures between 150 and 480 °C.

A common means of introducing the sample to the furnace and maintaining it in position in the centre of the carrier-gas flow is to use a boat attached to a rod. One end of the rod projects from the unit through a gas-tight seal or is actuated magnetically. The rod is pushed home so that the sample passes quickly from the cool inlet end of the pyrolyser into the centre of the furnace. Deur-Šiftar *et al.*[32] designed an example of this type (Figure 2.1).

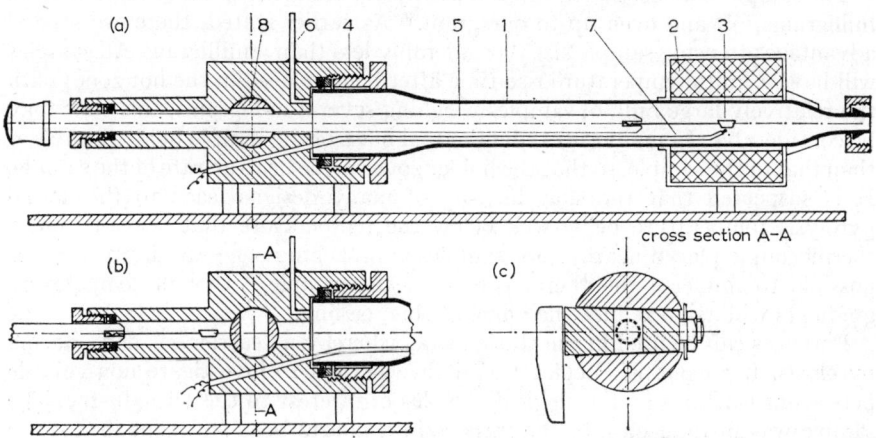

Fig. 2.1 The pyrolytic device designed by Deur-Šiftar and co-workers. (a) Complete device in operating position; (1) quartz tube, (2) electrical heater, (3) sample boat, (4) metal block, (5) pusher rod, (6) carrier gas inlet, (7) thermocouple, (8) teflon tap; (b) the connecting block, with the tap closed; (c) detail of the Teflon tap.
(Reproduced, with permission, from *J. Chromatog.*, 1966, **24**, p. 405)

The electrically heated furnace allows any pyrolysis temperature to be used up to its maximum, in this case 1000 °C. The temperature close to the pyrolysis point is maintained to $\pm$ 0.5 °C, as measured by a thermocouple. The temperature sensor needs to be close to the sample position, as the temperature difference between the wall and the carrier in the centre of the tube can be as high as 100 °C.[33] The sample is always brought to the same point for pyrolysis. The use of a boat means that no sample preparation is required, and accurately known amounts can be pyrolysed. Since heating is indirect, the boat should be of low thermal mass[34] and well insulated from the pusher rod; it should be made of material of low catalytic activity, quartz or noble metals often being used. In the design shown in Figure 2.1, further samples can be introduced without interrupting the flow of carrier gas by use of the Teflon tap. Gas flow through the chromatograph may also be maintained by using valves to by-pass the whole unit, but this necessarily complicates the path of the pyrolysis products, which is best kept as short and simple as possible so as to reduce the chance of condensation. Leplat[35] used heated electromagnetic valves to control the direction of flow of the carrier gas, allowing both dynamic and static flow modes to be utilised.

By placing the sample in a boat, contamination of the furnace is kept to a minimum, as the bulk of the residue will remain in the boat. However, there is always the possibility of spattering, which will cause contamination of the furnace walls and reduce the value of residue determinations made by weighing the tared boat after pyrolysis. A replaceable furnace liner[36] or a design that allows easy access to clean the furnace are of advantage.

The sample sizes used with furnace pyrolysers are often of the order of several milligrams,[37,38] and even up to one gram.[36] As earlier stated, there are several advantages to using sample sizes considerably less than a milligram. All samples will have a finite temperature rise-time after being placed in the hot zone; with the relatively large bulk of samples weighing several milligrams, this time may be considerable. In many cases, the thermal mass of the sample holder is greater than that of the sample, so that the holder governs the heat-up rate of the sample. It is suspected that the slow heat-up of many designs leads to the actual pyrolysis temperature being well below the temperature that is read from a thermocouple placed nearby. For samples approaching one gram in weight it is possible to immerse the thermocouple in the sample,[36] though temperature gradients will again reduce the value of this reading.

Furnaces can be used in the static mode, whereby no carrier gas flows during pyrolysis. For example, Leplat[35] used this method to investigate non-volatile petroleum fractions and geological samples of interest to the oil industry. The sample was introduced into the furnace in a quartz boat and kept in a static atmosphere for 5 minutes at 600 °C. The pyrolysate was then injected into the chromatograph over a period of 1 minute. This method can be expected to maximise secondary reactions, making a comparison of results from such a system with those from a flow-through system difficult or impossible. On the other hand, the secondary reactions may make discrimination of two similar samples possible where they appear identical with flow-through methods. Leplat preferred the static furnace method to the flow-through hot-filament method for his samples, and was able to obtain good reproducibility.

A static-mode pyrolyser has also been used in the g.c.–m.s. investigation of lunar samples, in which step-wise heating was employed. The sample was maintained at the required temperature for 5 minutes before the carrier gas was allowed to pass through the pyrolyser. The process was then repeated at a higher temperature.[31,39]

Pulse-mode Pyrolysers
Pulse-mode pyrolysers as a whole are probably more common than the continuous (furnace) type.[40,41] Of the range of types reported, only two are in common use, the hot-filament and the Curie-point types, and these will be described first.

(a) *Hot-filament Type*. In essence, a filament pyrolyser consists of a small coil of electrically resistive wire and electrical leads; the simplest adaptation will be made to fit into the injection port of the chromatographic column, common standard G.C. columns having i.d.'s of up to 3—4mm, which is adequate for the

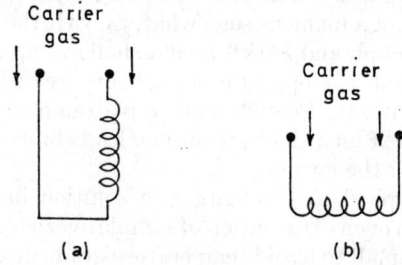

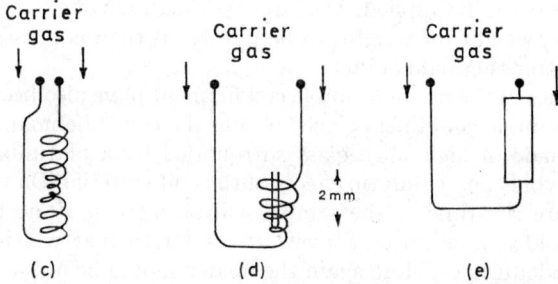

Fig. 2.2   Various designs of filament pyrolyser that have been used

insertion of a useful filament. The leads will leave the G.C. through a gas-tight seal to a source of power. Filaments with power and voltage requirements not greater than those of a car headlamp are common. For accuracy and flexibility, a variable timing device and variable voltage control are useful, enabling a range of accurate pyrolysis times and temperatures to be applied.

Filament pyrolysers have been in use for many years, and designs of various shapes and materials have been described. Jones and Moyles[15] described a nichrome filament with its axis parallel to the flow of carrier gas (Figure 2.2a). Jennings and Dimick[42] described a platinum coil with its axis at right angles to the flow of carrier gas (Figure 2.2b). It is doubtful that any significant difference in distribution of pyrolysis products is caused by such subtle design changes, though minor differences could occur.[19] Barlow, Lehrle, and Robb[43] made a nichrome filament narrowing to a 'basket' at the centre, so that samples could always be placed in exactly the same place on the filament (Figure 2.2c). This is important, as temperature gradients are bound to occur along the filament. Stanford[44] used an inverted cone-shaped filament to house a glass capillary containing the sample (Figure 2.2d). Dimbat and Eggertsen[45] used a platinum filament coated with glass to prevent hot spots due to shorting. Lehrle and Robb[46] used a nichrome ribbon filament, following the introduction of a foil filament that has a large surface area by Franc and Blaha[47] (Figure 2.2e). This is stretching the definition of a filament somewhat, as with the filament-cup[48] and filament-dish[49] types. Nelson and Kirk[48] used a shallow cup (1.5 × 3 mm) made of 0.001 inch platinum foil, the opposite sides being welded to platinum wires carrying the heating current. Perry[49] used a platinum dish about 5 mm by 4 mm and of material 0.028 inch thick; nichrome leads brazed at the mid-points of the short sides carried the current.

Samples are often applied to filaments as a solution in a suitable volatile solvent, thus ensuring an even distribution of sample over a region of the filament; this region is best kept small to avoid temperature gradients. The 'cup' or 'dish' pyrolysers have the advantage of easier sample preparation. In many cases the sample may be insoluble, and it can then be dropped onto the 'dish' direct; there is also no danger of interference by solvent peaks in the chromatogram, and weighed amounts can be applied. The larger surface area of a foil or dish allows a thinner sample, weight for weight, to be pyrolysed than is possible with a coil, thus giving better thermal contact.

The difficulties in the use of a simple coil filament have also been overcome by using various sample containers placed within the coil.[18] Lehmann and Brauer[50] used a boat made of high-silica glass surrounded by a platinum heating coil. This method avoids any catalytic effects of the coil material. On the other hand, the temperature rise-time of the sample will be increased due to the thermal mass of the holder, a situation somewhat similar to that which occurs in the continuous-mode furnace. Here again the holder should be of low thermal mass, for rapid heating, and a good flow of carrier gas should pass over the sample to minimise secondary reactions, if results obtained with pyrolysers of different design are to be compared. Lehmann and Brauer[50] used sample sizes of 2—3 mg [rather large for rapid and even temperature rise,[43] though smaller samples can be used with the technique (5—30 µg by Stanford, for example).[44]]

(i) *Temperature measurement.* The upper temperature limit achieved by a filament pyrolyser can be measured by the use of a thermocouple, as is commonly the case with continuous-mode furnace methods. The difficulties in accurately measuring this temperature by thermocouple are greater, however, as the temperature gradients are steeper than in a furnace, due to the relatively small heat source; placing is therefore even more critical. Lehmann and Brauer[50] used a thermocouple close to the sample, and they initially calibrated the system with a second thermocouple in contact with the sample. Since the upper temperature limit will only be reached at the filament itself, methods that measure the surface temperature of the filament will be more accurate. Barlow, Lehrle, and Robb[43] used a series of standard compounds of known melting point to calibrate the supply voltage to the filament temperature up to 800 °C. When 'glowing' temperatures are reached, a colour scale can be quoted as a rough estimate of filament temperature.[42] More precise measurements can be obtained with optical pyrometers, such as the disappearing-filament pyrometer; several workers have used such instruments.[15,43,51] The filament itself has been used as a resistance pyrometer.[52] Zulaica and Guiochon[51] calibrated their filament using three methods; (i) melting of salts, (ii) resistance of the platinum wire, and (iii) optical pyrometry. A systematic discrepancy of about 30 °C was found. The upper temperature limit is best measured by a method giving the temperature of the filament at the point where the sample is to be placed; salt melting points or optical methods involving focussing on that point should be better in this respect than a resistance method that measures resistance through the full cross-section and along the length of the filament. Recently, a low-mass thermocouple has been used that was welded to the filament at the sample point.[53] If temperature readings are taken from a calibration curve of filament temperature against current or voltage, the curve must be checked from time to time to allow for changes in the resistance of the filament. Jennings and Dimick[42] found that a considerable increase in current was required after twenty firings to maintain the temperature. This was due to carbon dissolving in the filament, which, incidentally, in so doing also embrittled the coil. The number of firings achieved before this effect becomes noticeable will vary considerably, depending on the filament temperature, material, and sample type.

(*ii*) *Heating rate*. It was mentioned earlier in this chapter that the time taken by the sample to reach the upper temperature limit is an important factor in determining the composition of the final mixture of pyrolysis products if a significant proportion of sample is pyrolysed during the heating period. By keeping the sample size low, and its area in contact with the heating source high (as with ribbon-type filaments), the temperature rise-time in the case of pulse-mode units (such as filaments) is a function of the design of the unit. The upper temperature reached by a filament using resistive heating is determined by the equilibrium between heat input (a function of the product of the square of the current and the resistance of the filament) and the heat loss to the carrier gas, radiation to the walls, and conduction through the leads. A low thermal mass of the heated system will ensure a high initial heating rate but the equilibrium temperature will be approached asymptotically, *i.e.* the heating rate must slow near the upper temperature.

(*iii*) *Boosted filament units*. Several methods have been used to overcome the relatively low heating rate of a filament supplied with a constant voltage. Krejci and Deml[52] used a variable voltage source, the upper temperature being controlled by constant monitoring of the filament resistance by incorporating the filament in one arm of a Wheatstone bridge (temperature rise-time was 80 ms). Cogliano used a short high-voltage boost to attain pyrolysis temperature.[54] Lehrle and Robb[46] incorporated the filament in a bridge circuit and produced a diagram (Figure 2.3a) showing the long heat-up time of the filament. A boost current was then applied for one second and adjusted to give the heat-up profile shown in Figure 2.3b. This principle has been further developed and commercially utilised. Levy[13] obtained temperature rise-times of 15 ms, using a voltage sweep

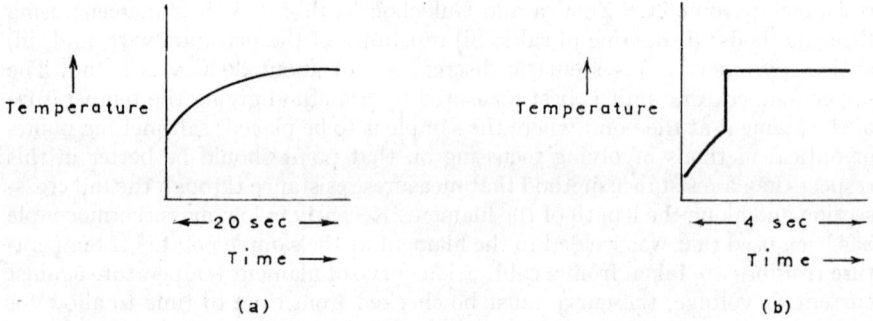

Fig. 2.3   Diagram showing the temperature rise with time of (a) an unboosted filament, (b) a boosted filament pyrolyser
(Reproduced, with permission, from *J. Gas Chromatog.*, 1967, **5**, p. 89)

obtained from a capacitor-discharge power supply; the temperature was measured optically, using a photomultiplier and recording the output on an oscilloscope, and more recently 700 °C has been reached in 12 ms.[53] The sample thickness must be very small if one is to take advantage of these higher heating rates. Lehrle and Robb[46] showed that the degradation rate of poly(methyl methacrylate) became independent of sample thickness at thicknesses below 400 Å. Sample sizes down to 10^{-8} g were used on a ribbon filament to reduce the sample thickness. The filament's upper temperature is probably a good enough approximation of the pyrolysis temperature for most purposes, but Levy, Fanter, and Wolf[53] have defined the true pyrolysis temperature as 'that temperature at which the rate of energy (power) consumed by the sample is equal to the net power supplied to the system', which can be seen as a plateau in an oscillo-graph trace of the temperature *versus* time curve during heat-up.

(*b*)   *Curie-point Pyrolysers.* Induction-heating devices were introduced to p.g.c. by Szymanski, Salinas, and Kwitowski,[55] who mixed the sample with iron filings or other ferromagnetic material. Andrew, Phillips, and Semlyen[56] showed the similarity of the pyrolysis products obtained in this way with those from a filament pyrolyser used by Janák.[57]

Induction heating became a technique rivalling hot filaments in advantages with the design of Simon and Giacobbo.[58] This gives a fast heating rate[59] up to the well-defined limit of the Curie point. This latter effect distinguishes it from the previous induction heaters. In the Curie-point pyrolyser a ferromagnetic wire is centred in a glass or quartz tube, which is connected to the inlet of a gas chromatograph, and through which the carrier gas flows. A high-frequency induction coil surrounds the tube and heats the wire by induction. The wire heats up till its Curie point is reached (this is the temperature at which the wire becomes paramagnetic, and its energy intake drops, thus holding the temperature of the wire at this point). Different pyrolysis temperatures are obtained by using wires of different Curie point. A range of temperatures is obtained by using alloys containing differing amounts of the common ferromagnetic metals iron, cobalt, and nickel. Since the Curie point represents a considerable change in the electromagnetic properties of the pyrolysis wire, oscillators delivering high power inputs at temperatures below the Curie point can be used without danger of overheating the wire. Nevertheless, the Curie point does not represent an instant change from ferromagnetism to paramagnetism, but the temperature at which ferromagnetism is finally lost; hence the heat-up cannot be instantaneous.[53] The temperature rise-time is still very rapid if the oscillator output is sufficient; Simon and Giacobbo used a 1.5 kW supply. Some factors affecting temperature rise-time and final temperature have been investigated by Bühler and Simon[60] and by Levy, Fanter, and Wolf.[53] Oscillators with outputs similar to the power used in a resistive filament will give longer temperature rise-times.[13] The upper limit of power input is dependent on a number of factors, including wire dia-meter and oscillator frequency.[60] Temperature rise-times are dependent not only on these factors but also on the geometry of the induction coil (which influences the magnetic field strength) and on the nature of the elements present in the wire. Oscillators producing a few tens of watts to several kilowatts[13] (2.5 kW) have been used.

With a correct choice of components and couplings,[50] temperature rise-times comparable with those of boosted-current filaments are achieved, with the added advantage of well-defined upper temperatures. All wires of given composi-tion and dimensions will behave similarly,[60] and can therefore be replaced for each new pyrolysis without recalibration. Reproducibility is said to be affected by changes in the surface, as seen by electron microscopy.[60] There is also a danger that ferromagnetic material (with a high Curie point) contained in the sample in sufficient quantities could cause the sample to heat up beyond the Curie point of the wire.

As a sample holder, a Curie-point wire is more versatile than a filament, be-cause it can be preshaped to suit individual samples[61] (within limits, that is) as coiling can slow the heat-up rate.[53] Most of the drawbacks mentioned above will not affect the great majority of analytical applications, but there is no way of using a Curie-point pyrolyser in a stepped-temperature pyrolysis, in which one sample is subjected to a series of pyrolyses at increasing temperatures.

(c) *Dielectric-breakdown Pyrolyser.* Barlow, Lehrle, and Robb[43] designed a dielectric-breakdown pyrolyser (Figure 2.4) in which the sample acted as the dielectric of a condenser across which voltages up to 3 kV were applied. A small triggering voltage, which did not noticeably contribute to the production of pyrolysis products, was applied and a known energy was then discharged through the sample from a capacitor *via* a series resistor controlling the discharge time. Temperatures of many thousands of degrees were believed to be

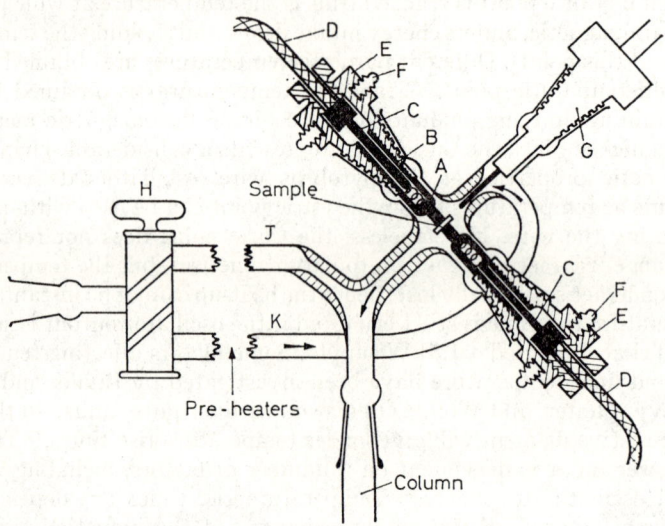

Fig. 2.4 Diagram showing the position of the sample in a dielectric-breakdown pyrolyser. A—F is the suspension and electrical connections, G and H are isolating taps.
(Reproduced, with permission, from *Polymer*, 1961, **2**, p. 30)

attained in the sample, which afterwards showed a small hole, with the loss of a few tenths of a milligram in weight. Not surprisingly, the major products from a series of polymer samples were of low molecular weight, and mainly unsaturated hydrocarbons. Even so, some styrene and methyl methacrylate, for example, were detected from the breakdown of their respective homopolymers, though in other cases (*e.g.* polyethylene) characteristic pyrograms were not obtained. The method was therefore considered by Barlow *et al.* to be of use only for those polymers which do not degrade entirely to small fragments under the drastic conditions produced in this apparatus.

(*d*) *Electric Arc Pyrolysers.* Johns and Morris[62] placed electrodes through the sides of their pyrolyser so that an arc could be struck at right angles to the flow of carrier gas. A sample holder positioned the sample in the arc, and the arc (passing 55 mA at up to 5000V) was maintained for 20s. The pyrolysate was taken into a sample loop before being transferred to the G.C. Levy[18] has pointed out that this system gave a pyrogram for polyethylene considerably different from the hydrocarbon series obtained by other workers with different pyrolysis units, and he concluded that the pyrolysis conditions differ considerably from the conditions in conventional units.

A modification of this system was described by Sternberg and Litle.[63] The electrodes were placed along the axis of the flow-through discharge tube. The sample was placed on the downstream electrode, which was made of porous graphite felt; this ensured rapid removal of pyrolysate from the arc. Again the pyrolysate was first passed into a sample loop. The apparatus allows the possibility of pyrolysis of liquids and gases adsorbed on solid supports, though there

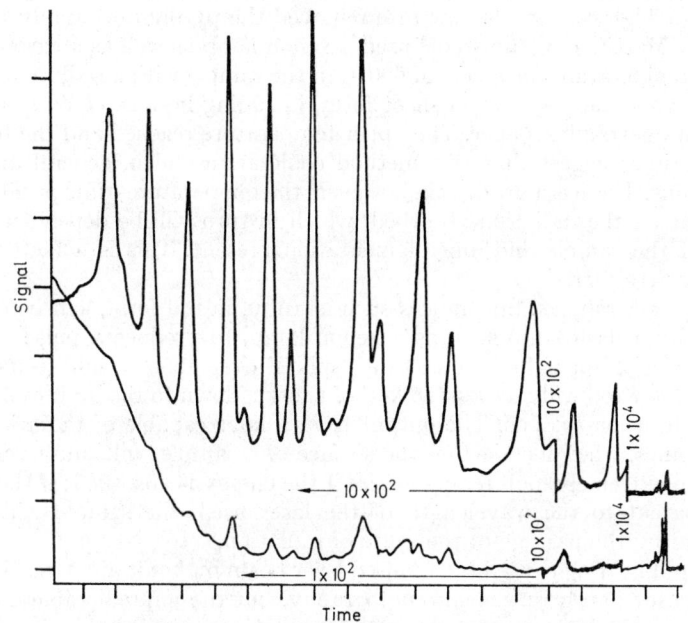

Fig. 2.5 Pyrograms showing the effect of sample position on the fragmentation of polyethylene samples. Upper curve: sample near downstream electrode. Lower curve: sample near upstream electrode.
(Reproduced, with permission, from *Analyt. Chem.*, 1966, **38**, p. 324)

will be a greater danger of catalytic effects due to the supports. Pyrograms of polyethylene, pyrolysed near the upstream electrode and near the downstream electrode, showed the importance of downstream pyrolysis if characteristic pyrograms are to be achieved (Figure 2.5). With a more sophisticated version of the set-up than that which produced the ones shown, pyrograms resembling those produced by a filament pyrolyser are obtained. Sternberg and Litle estimated their pyrolyses to be equivalent to the 800—850 °C pyrolyses obtained by Lehmann and Brauer[50] in their filament pyrolyser, which used a silica boat as sample holder. This temperature was indicated by a comparison of pyrograms obtained from poly(methyl methacrylate) and is low enough to make the technique useful in most analyses by pyrolysis.

These pyrolysers are distinguished from that described below in that the sample is contained within the electric arc.

(e) *Thermal Radiation Pyrolysers.*   (i) *Carbon arc and xenon lamp radiation.* Sheet cellulose was pyrolysed by Martin[64] by exposing the sample 'less than one inch in diameter' to uniform radiation from a carbon arc. The sample was supported in the stream of carrier gas and the radiation introduced through a quartz window. A high carrier-gas flow rate removed the pyrolysate rapidly from the radiation. Martin and Ramstad[65] used a xenon lamp as well as a carbon arc to produce temperatures in excess of 600 °C in the sample with a radiation pulse of duration 1 ms. Samples not in sheet form, including liquids, could first be absorbed on quartz-fibre paper. The upper temperature reached and the temperature rise-time suggest that this method could be useful in general analytical applications. There is a drawback, however; the temperature–time profile will be dependent on the radiation absorbed, which in turn will be dependent on the nature of the sample and any pigmentation present. This is a factor in any radiation pyrolyser.

(ii) *Laser pyrolysers.* Substances such as ruby, in rod form, can be optically pumped by a flash lamp such as a xenon lamp, to produce a pulse of monochromatic coherent light in a series of 'spikes' over about a millisecond. A Q-switched laser rod will give less 'spikes' of energy, down to the limit of one giant spike of duration *ca.* 1 ns. The output from a laser capable of radiating up to several joules, when focussed on the surface of a sample, will cause very rapid heating to extremely high temperatures if the energy is absorbed; if the sample is transparent to the wavelength of the laser used, the sample will remain unaffected by the passage of the pulse of radiation.

The rapid heating caused by a pulsed laser beam makes it an attractive heating source for pyrolysis–gas chromatography, but the unpredictable and probably excessively high upper temperatures cause serious problems in achieving characteristic and reproducible pyrograms.

Wiley and Veeravagu[66] found mainly methane and acetylene in the vapour from capillary tubes containing aromatic hydrocarbons subjected to light from a ruby laser, suggesting that there was severe breakdown of the sample. The irradiations were carried out in a static atmosphere, which was subsequently sampled for g.c. analysis. Folmer and Azzaraga[67] used a ruby laser focussed onto a sample placed on a boat in the stream of carrier gas. They reported simple but characteristic pyrograms from this arrangement for a variety of substances, and used a mixture of the sample ground with graphite to cause absorption where the sample was transparent to the radiation.

Guran, O'Brien, and Anderson[68] produced no styrene monomer from poly-(styrene–butadiene) unless the i.r. beam of the neodymium-in-glass laser rod was deliberately defocussed; this suggests that the temperature reached in the focussed beam is too high for most analytical purposes. Contrary to the views of Guran *et al.* on the reproducibility of the technique, it is the opinion of the authors of this article that the presence of pigments reduces the efficacy of the technique, as crater size is strongly dependent on the colour of the sample; this suggests that pyrolysis temperature must also be dependent on the colour.[69] Ristau and Vanderborgh[70] used a quartz or Vycor tube of circular cross-section as the pyrolysis chamber and 'focussed' the output from a laser through the curved surface. The tube was replaced from time to time because fractures were caused by the laser beam. With this set-up, styrene was obtained from polystyrene, but only to the extent of 23% of the detected product.[71]

Graphite or other carbonaceous material has been used by several workers[67,68,72–74] to increase the absorption of radiation. Folmer[74] showed that the carbon can lead to changes in the pyrograms compared with unaltered samples, reproducibility being dependent on using a fixed proportion of graphite with the sample, though coating the sample on the beam-exit side with graphite or gold did not cause changes in the pyrogram. Ristau and Vanderborgh[71] avoided the use of carbon as an absorber of laser energy where possible because of the discrepancies in the pyrograms that they observed. Fanter, Levy, and Wolf[75] used a sample holder of blue glass (Figure 2.6) as a 'back stop' for the radiation, and hence as a secondary energy source to avoid the interference with the pyrogram caused by admixture of graphite. They concluded from their experiments with polystyrene and polyethylene at different power output levels from a ruby laser that the product distributions are similar to those for conventional thermal pyrolyses at 1200—1500 K, much lower than the temperatures expected in a sample absorbing a pulsed laser beam. The lower 'equivalent' temperatures compared with conventional electrically heated pyrolysers could in part be due to lower chances of secondary reactions.[67] The 'equivalent' temperatures reported[75] (though low by pulsed laser standards, where temperatures of several thousand degrees are postulated[76,77]) are still well above the pyrolysis temperatures normally required for fingerprint pyrolyses.

Lack of reproducibility due to variable absorption from sample to sample, difficulties in achieving good focus, and the high pyrolysis temperatures make the laser pyrolyser of limited use. One application has been found in the evaluation of the hydrocarbon content of oil shales,[78] using a focussed, pulsed ruby laser. In this design, the quartz window through which the beam is focussed is separated from the sample by a six-inch tube, to prevent fogging of the window by pyrolysis products; the dead space of this tube is cut off from the carrier-gas flow by an electromagnetic valve a few milliseconds after the laser shot.[79a] Further applications are found where a small precise area must be analysed.[79b]

Others
A radiation-breakdown technique, photolysis–gas chromatography, has recently been described, in which breakdown is achieved over a period of 5—30 min by u.v. radiation. The technique as described[80] will require more development before it is likely to rival p.g.c. for most analytical purposes.

A furnace-type pyrolyser designed as part of a possible system for the automatic biological exploration of Mars has been described.[81] As with the pyrolyser used for samples of meteoric and lunar origin,[30,31,39] it is designed to take large (15 mg) samples consisting largely of inorganic matter. Unlike the vast majority of furnaces, this type operates in the pulse mode, the sample being heated from ambient temperature to 500 °C in 15 s, a very slow heat-up rate; clearly, this apparatus is for very specialised use.

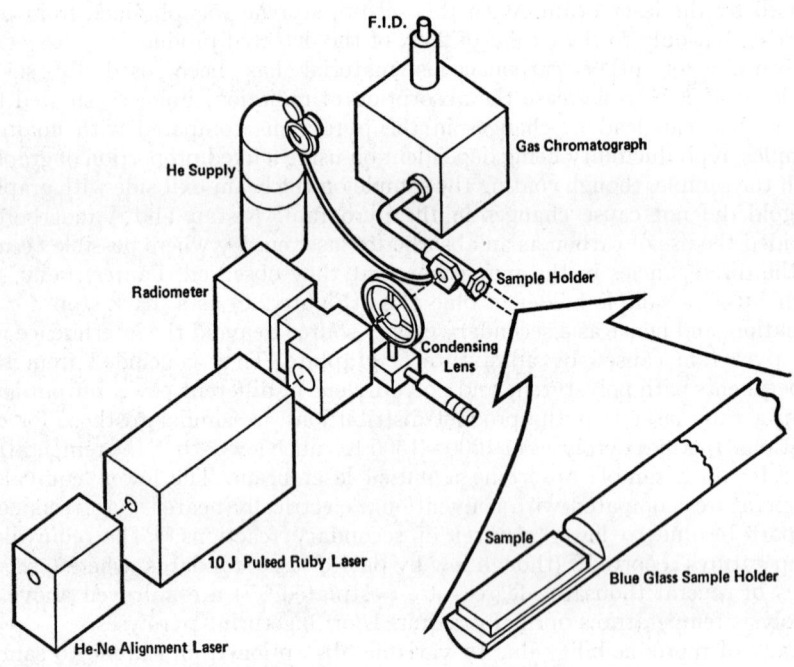

Fig. 2.6 Schematic diagram of system for laser pyrolysis and analysis. The special blue-glass sample holder is shown enlarged
(Reproduced, with permission, from *Analyt. Chem.*, 1972, **44**, p. 44)

Apparatus for the Pyrolysis of Gases and Volatile Liquids

Volatile liquids which vaporise at temperatures below the required pyrolysis temperature can be treated as gases, and pyrolysed in a system designed for vapours (usually a form of furnace).

With a through (flow) reactor, the residence time in the hot zone will be controlled by the carrier-gas flow rate and the internal dimensions of the reactor. There is no sample holder, so that the heating rate will be similar to that of the carrier gas if the sample does not form too large a fraction of the volume in the pyrolyser. Faster heating rates are obtained if the carrier gas is preheated. As with solids, heated injection ports have been used as low-temperature pyroly-sers, for example in producing fatty acid methyl esters from tetramethylam-monium salts,[82] and many purpose-designed pyrolysers have also been described.

A simple tubular furnace capable of withstanding high temperatures was designed by Keulemans and Perry[83] (Figure 2.7) consisting of a quartz tube of

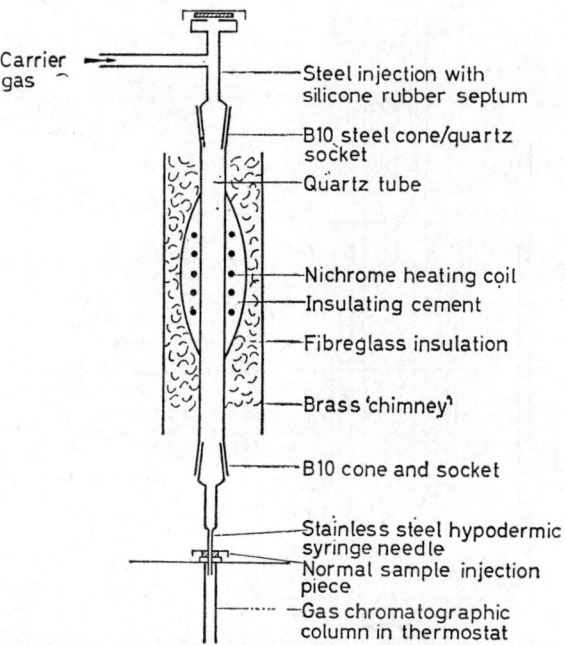

Fig. 2.7   A quartz reactor used for thermal decomposi-tion of samples
(Reproduced, with permission, from 'Gas Chromato-graphy 1962', ed. M. van Swaay, Butterworths, London, 1962, p. 358)

10mm i.d., with the centre 14cm heated to about 500 °C by a nichrome coil. The carrier gas was preheated and passed through the reactor at the equivalent of 25 cc min^{-1} at S.T.P., so that the residence time in the cracker was 20 s. Tempera-tures were measured by thermocouple at the centre of the heated zone. A series of hydrocarbons up to C_{10} were pyrolysed reproducibly in this apparatus. The use of temperatures below 550 °C and liquid samples of less than 1 μl ensured that secondary reactions were negligible. Catalytic activity was minimised by using an open quartz tube.

Two concentric stainless-steel tubes formed the basis of a simple pyrolyser used over the range 200—1000 °C to fragment a range of simple organic com-pounds.[84]

A sophisticated design, giving accurate control of temperature and reaction time, was described by Cramers and Keulemans[85] (Figure 2.8). This consisted of a one metre reactor tube of i.d. 1 mm, made of copper, silver, or gold, the last being preferred. A core and heated jacket of silver enclosed the coiled tube to

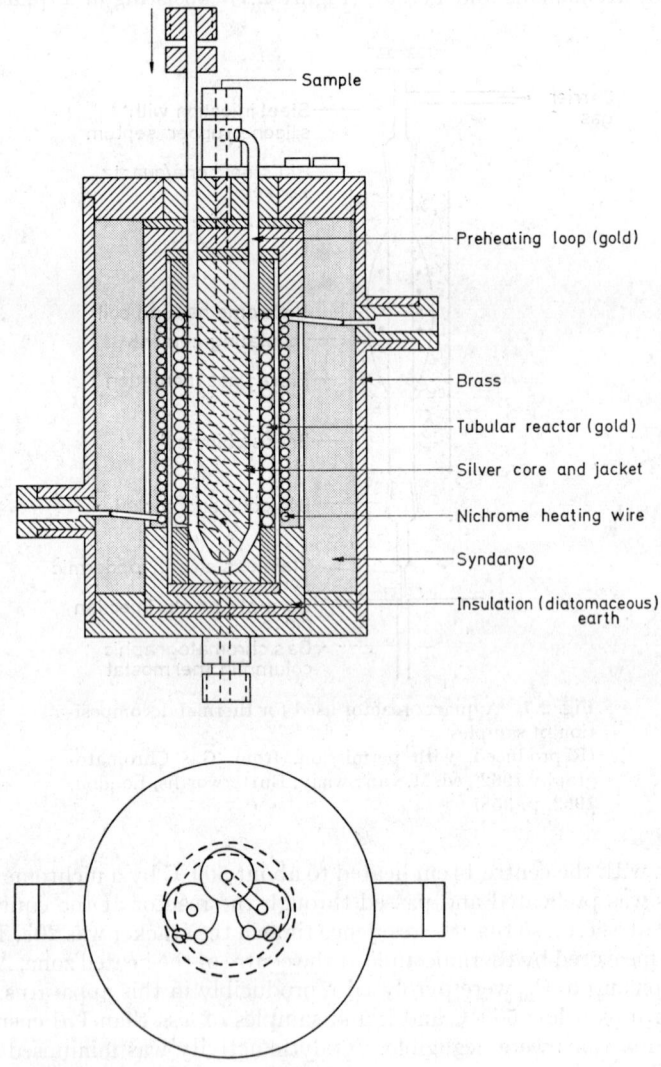

Fig. 2.8   Diagram of a pyrolyser that allows accurate control of temperature and reaction time
(Reproduced, with permission, from *J. Gas Chromatog.*, 1967, **5**, p. 59)

give good thermal conductivity and temperature homogeneity, and also provided pre-heating for the carrier gas. The narrow-bore tubing improved thermal contact and hence increased heat-up and cool-down rates at the extremities of the apparatus, but the larger surface area must increase the chances of catalytic reactions. (Stainless-steel and platinum reactor tubes were found to give different pyrograms from tubes of the above materials.) Gold tube pyrolysers have been used recently by other workers.[86-88] Packed furnaces have been employed to increase the heating rate of the entering gases, the packing acting as a heat sink. Glass or quartz wool and beads,[89] gas chromatography support materials,[90] and acid-washed sand[91] have been used as packing materials. The use of packed

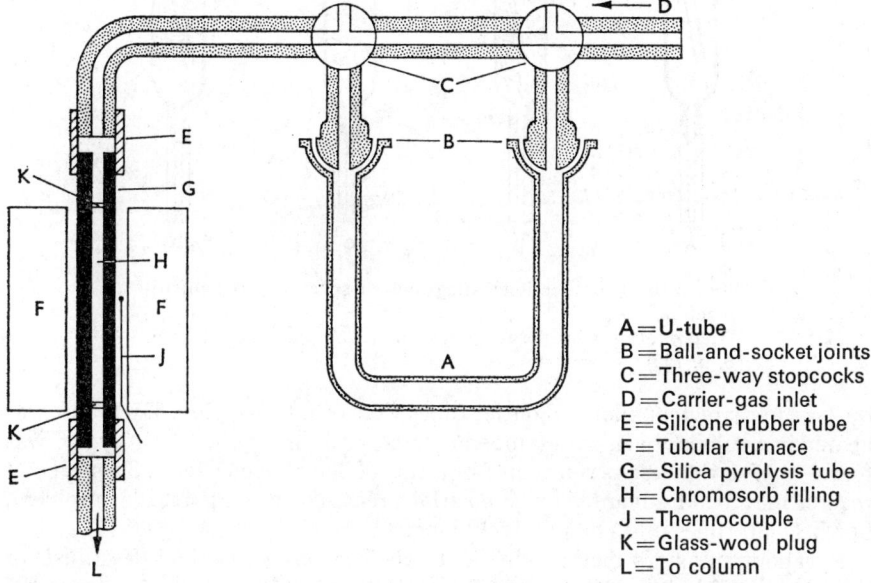

A = U-tube
B = Ball-and-socket joints
C = Three-way stopcocks
D = Carrier-gas inlet
E = Silicone rubber tube
F = Tubular furnace
G = Silica pyrolysis tube
H = Chromosorb filling
J = Thermocouple
K = Glass-wool plug
L = To column

Fig. 2.9   Diagram of an injection system and pyrolysis unit designed by Dhont (Reproduced from *Analyst*, 1964, **89**, p. 72)

columns is likely to increase the proportion of pyrolysate produced by surface catalysis. Figure 2.9 shows a design by Dhont,[92] consisting of a tube 20 cm long and of 2 mm i.d., heated over a 13 cm length by a furnace and packed for 10 cm of the heated length with Chromosorb P. The reactor was used with 50—500 mg of vapour samples from the U-tube gas-sampling device which is also shown in the diagram. It may prove possible to pyrolyse adsorbed gases retained on charcoal or support materials in the recently available commercial equipment referred to in the section on boosted filament units.[104]

A unit for the pyrolysis of aqueous solutions, consisting of a nickel-packed nickel tube, has been described;[119] for use with p.g.c., packings that are less likely to be catalysts would be desirable. A series of four Pyrex microreactors with varied internal structure have been examined as possible reactors in kinetic experiments[93] (see Figure 2.10). A reactor with low surface : volume ratio was sought without introducing excessive dead volume. The reactors were 5 cm long

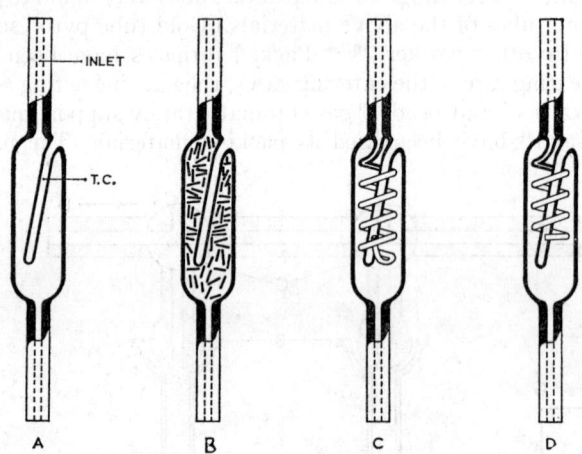

Fig. 2.10   Cross-sectional diagrams of several Pyrex micro-reactors
(Reproduced, with permission, from *J. Chromatog. Sci.*, 1970, **8**, p. 722)

by 1 cm in diameter; inlet and outlet tubes were of i.d. 1—2 mm. Each unit was heated in a tubular oven with temperature control better than 1°C. Type C was found to give the best performance in terms of the above criteria. This type of reactor may be of advantage where surface reactions are a particular problem, but heat-up times must be relatively long.

A vapour pyrolysis method similar to the carbon arc method described for solid samples[63] has been developed.[94] An a.c. discharge was passed through the carrier-gas stream between platinum electrodes to cause fragmentation of vapour samples injected into the carrier gas. The effect of varying current and the composition of the carrier gas was investigated with a number of volatile compounds. A problem (common to all vapour pyrolysers discussed) is that characteristic pyrograms are usually obtained only at low (about 20%) conversion of volatile samples; at higher proportions of breakdowns, smaller, less characteristic breakdown products are formed,[94] and the lower limit of detector response may necessitate the use of relatively large samples.

Goforth and Harris[95] discussed the design features of vapour pyrolysers in detail, and concluded that a long tube of small diameter taking a high carrier-gas flow rate minimised heat-up and cool-down times. They preferred quartz as the tube material owing to its low catalytic activity, which they believe out-weighs its low thermal conductivity when compared with metals.

Static-mode pyrolysers have been used for vapour samples in special cases such as the low-temperature (50—180 °C) pyrolyses of diboranes[96] in a vapour-jacketed vessel. The remarks about secondary reactions made on the subject of static pyrolysis of solids will apply equally to vapours.

Tandem G.C. Systems

The main types of gas-phase pyrolysis apparatus are described in the previous section of this chapter. In theory, any such apparatus can be coupled between two G.C.s to form a G.C.–P.G.C. system to analyse eluates from a G.C. without an off-line trapping stage. Such complete systems will be described in this section.

Keulemans and Perry[83] suggested that the major requirements of a tandem system are (a) the initial (separation) column should be of high separation

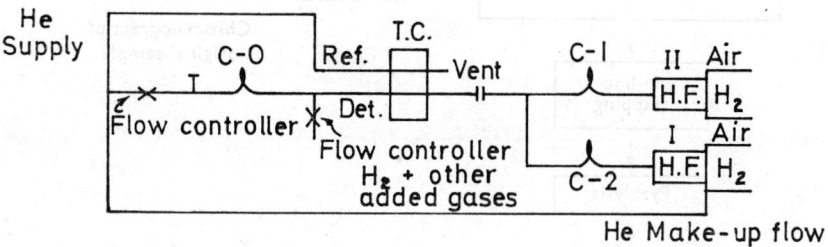

Fig. 2.11   Column configuration for a tandem G.C. system. Column 0: 8′ × ¼″ silicone (Thermotrac I); column 1: 8′ × ⅛″ Apiezon; column 2: 8′ × 1/16″ alumina (Thermotrac II) (Reproduced, with permission, from *Analyt. Chem.*, 1966, **38**, p. 1640)

efficiency and (b) the second, fragment-separating (analysis) column should be temperature-programmed so as to elute the higher-boiling fragments and the undecomposed parent compound rapidly. This second column should have a non-selective stationary phase, so that the approximate boiling point of the parent can be obtained.

A system was described by Sternberg *et al.*[94] (Figure 2.11) in which there was an initial column containing silicone, C-0, the effluent from which passed through a thermal-conductivity detector before passing on to an electric-discharge pyrolyser. After pyrolysis, a stream splitter passed a portion of the pyrolysate to a column containing Apiezon grease as the stationary phase. The rest passed to a column containing alumina, which was temperature-programmed to elute hydrocarbons up to carbon-number six. The system as described could only be used to pick out a single eluate for pyrolysis, unless conditions in each column were chosen so that the columns C-1 and C-2 could complete the analysis of each pyrolysate before the next eluate from C-0 entered the pyrolyser. The G.C. columns and conditions are not sufficiently well defined in the published work to permit reproduction of the data.

Goforth and Harris[95] used an in-line trapping system to delay eluates from the separation column before passing them on in short 'plugs' to the pyrolyser

(Figure 2.12). No handling or transfer of samples was required. A second stream of carrier gas entered the line ahead of the trap to increase the flow and hence to reduce retention times in the analytical column, and also to allow optimisation of flow rates to each column. The single trap serves only to concentrate the successive peaks from the separating column. The analysing

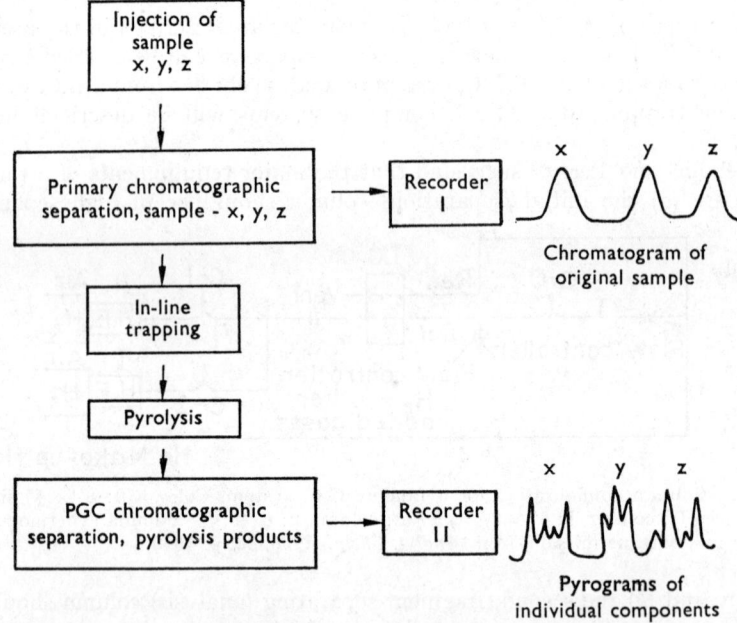

Fig. 2.12   Schematic representation of the system for the identification of gas chromatographic peaks by p.g.c.
(Reproduced, with permission, from 'Gas Chromatography 1968', ed. C. L. A. Harbourn, Institute of Petroleum, London, 1969, p. 262)

column must still, therefore, elute each pyrolysate in the time between elution of successive peaks from the separating column, as in the previously described system. The analysing column was therefore heated in a rapid temperature programme at heating rates of up to 400 °C per minute; this was achieved by direct resistive heating of the metal column. The initial, separating, column must be chosen for each separation so that retention times of the components of the mixture differ by at least the 'turn-around time' of the subsequent analysing system. This latter time was very short (minimum 105 s). However, one would expect that the analysing column must be of limited performance in terms of resolution, due to the very short time allowed for analysis.

Levy and Paul[97,98] described a system containing three separate streams of carrier gas, interconnected by multi-port valves. The first system contains the separating G.C., the second contains the pyrolyser and a delay coil, and the third the analysing G.C. (Figure 2.13). By allowing three separate flow systems, each can have its optimum flow conditions, and a few selected peaks can be directed through the pyrolyser and then the analysing G.C., other peaks being vented to atmosphere. In this way, the analysis section can be run without a time restriction. Non-destructive detectors can be used to indicate the passage of each selected peak through the system. By splitting the pyrolysate stream into two, passing a portion through a 'small molecule', highly retentive, column and the rest through a 'large molecule' column, a considerable amount of information about the identity of the original eluate peak can be obtained.[87,99]

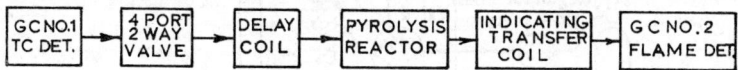

Fig. 2.13   Block diagram of the G.C.–P.G.C. system used by Levy and Paul (Reproduced, with permission, from *J. Gas Chromatog.*, 1967, **5**, p. 137)

Walker and Wolf[100-102] described a system incorporating interrupted elution in the separation G.C. (Figure 2.14). The four-port valve prior to the separation G.C. allows the carrier gas to pass through the separator G.C. until a component is detected by detector D_S, which takes only 10% of the flow from splitter S. The four-port valve is then rotated so that the carrier gas flows through the lower dotted track in the valve, isolating the separation G.C., which is then in the stopped condition. The separated component then continues through the pyrolyser and the pyrolysate is analysed without a time restriction imposed by continuous operation of the separation G.C. After the complete elution of the pyrolysate plus parent compound from the analysis G.C., the four-port valve is returned to its original position to re-start the separation G.C. until the next peak of interest is detected by D_S. The splitter in the analysis G.C. at D_A allows 90% of the flow to by-pass D_A and hence to be available for other analytical procedures. The separating column is chosen to suit each individual separation and can be temperature-programmed. Without a trapping stage, it is necessary that sharp peaks be produced so as to present a short plug of material to the pyrolyser, so that resolution in the analysing column is not reduced. This is an important consideration in choosing the separating column and its temperature programme. The choice of analysing column is said to require careful consideration, at least partly influenced by the nature of the original mixture. For hydrocarbons, a 25% DC 200 liquid phase on 60–80 mesh Chromosorb P was used in a 15 ft stainless-steel tube of $\frac{1}{8}$ inch o.d. held at 170 °C. With extra valving between pyrolyser and analysis G.C., the latter can be flow-programmed to reduce analysis time. This system allows the analysis of all well-separated eluates from the separation G.C.

A combination of the interrupted-elution and delay-loop methods was used by Merrit and DiPietro.[103] The delay loop and pyrolyser unit was a commercial persion of the system of Levy et al.,[18] and included a column packed with vorous polymer for separation of small molecules.

The systems described above have mainly been used for analysis of hydro-carbons and for functional groups. In the former case the use of different analysis columns and a variety of column conditions shows that standardisation in even this limited field has received little consideration. In the analysis of

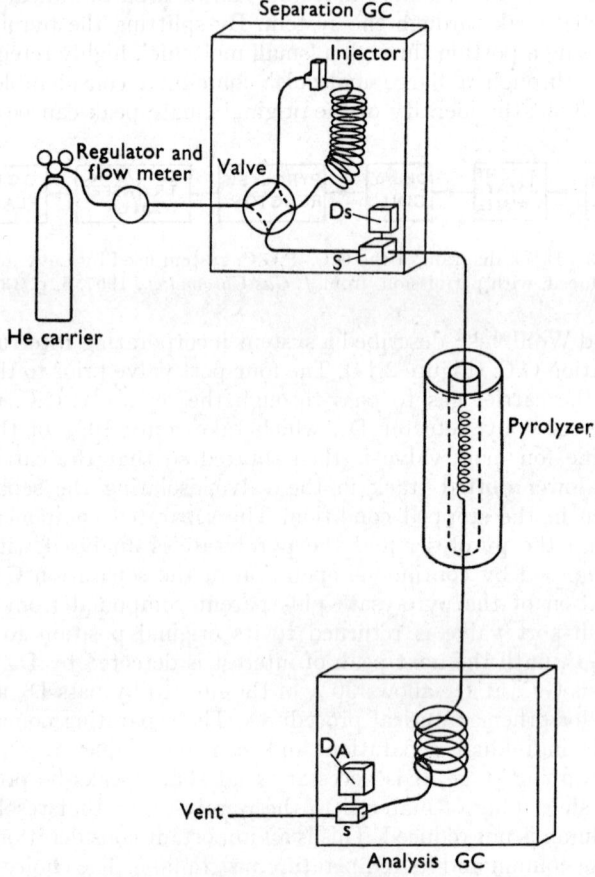

Fig. 2.14 Block diagram of complete analysis system employing interrupted elution and p.g.c.
(Reproduced, with permission, from 'Gas Chromatography 1968', ed. C. L. A' Harbourn, Institute of Petroleum, London 1969, p. 387)

functional groups involving the estimation of 'small molecules', the problem with respect to the analysis column is less severe. If agreement on those 'small molecules' which need to be quantified to characterise each functional group can be reached, then any column satisfactorily separating these compounds could be used, having been calibrated by injection of the 'small molecules' themselves. This still leaves the problem of standardisation of the pyrolysis system and conditions. Since quantification of most of the 'small molecules' is necessary, the conditions would have to be precisely defined before inter-laboratory comparisons could be made. A useful data collection could then be formed, as with the standardisation of polymer fingerprinting. Without this data collection, which is dependent on standardisation of conditions, the technique (though cheap in terms of equipment costs) compares unfavourably with mass spectrometry.

Comparative Studies of Available Techniques

A high proportion of the many different pyrolyser types described in this chapter are commercially available. No single type will prove best in all applications, so that each type must be considered in the light of the particular problem to be tackled. Some of the favourable points and drawbacks to the various techniques have been discussed in this chapter, and some comparative studies of different pyrolyser types have been published. Cost of apparatus, instrument operating time, and ease of sample preparation all affect operating costs, but technical requirements may over-ride these considerations. For example, a qualitative comparison of pyrograms produced in rapid succession may be all that is required; on the other hand, quantitative comparisons with results from other laboratories may be necessary. The possibility that others may wish to use the results for comparison is a question often given insufficient consideration. Factors such as consistency of temperature rise-time are important in some analyses but variations have little effect on others.[105]

The comparative studies published to date have had limited aims and therefore have not considered all the factors mentioned above, but are none-theless of considerable interest. Early work on dielectric-breakdown and hot-filament pyrolysis of certain polymers showed the filament technique to be of more general application.[43] Laser pyrolysis was compared with both filament and furnace methods by Folmer and Azzaraga;[67] Jones and Reynolds[19] have commented on the lack of information on sample size in this study. Under the conditions used, the laser system gave the simplest and most characteristic pyrograms. (G.c. conditions were slightly different for the laser pyrolyses.) It would be interesting to know if the results would have remained as favourable to the laser system if lower temperatures had been used in the filament and furnace methods (temperatures quoted were 1273 K and 1073 K). Tukeda and Mushiko[106] compared the use of a stainless-steel reactor heated to 600–750 °C with a high-voltage electric discharge for the pyrolysis of a wide range of volatile organic compounds. Under the chosen conditions the discharge method was found to be more suitable for the identification of the compounds.

In comparisons of more common types of apparatus, the following results have been obtained. Two model compounds, 2,6,10- and 2,6,11-trimethyl-dodecane, were pyrolysed by three different pyrolysers: (1) a capacitative boosted filament heater; (2) a high-power (2.5 kW) Curie-point heater; (3) a vapour-phase flow-through gold tube reactor. The results were compared.[107]

The pyrolysers were each operated at precise temperatures near 600 °C. A theoretical product distribution was also calculated for each compound at 600 °C, based on the free-radical theory of Kossiakoff and Rice.[108] The three methods were also compared over a range of temperatures. The sample size was kept to 0.6 μl in the case of the tubular pyrolyser and 0.5 μl for the Curie-point apparatus (the filament was dipped in the sample, and quantities were not quoted). The results show marked divergencies in product distribution between the methods and between each method and the theoretical value. The quoted stoicheiometry of the reactions showed a large variation in percentage decomposition from method to method. The Curie-point system caused only about 4% decomposition, the filament 7—30%, and the tube pyrolyser 13—94%, depending on temperature and sample. Clearly, the percentage decomposition was much lower for the Curie-point and filament methods than is expected for involatile samples;[109] the temperature rise-time was 120 ms for the Curie-point and 15 ms for the filament pyrolyser; the volatility of the samples used may have exaggerated the importance of the differences in these temperature rise-times compared with the results to be expected with involatile samples. Van Cauwenberghe *et al.*, [110] in a study of the pyrolysis of alkylbenzenes, concluded after their experiment that the substances investigated were indeed too volatile for use with the Curie-point technique, and they used a furnace filled with quartz wool in preference.

When using samples more appropriate to filament and Curie-point systems, namely a series of polyisoprenes, Galvin-Vacherot[111] found no differences in the nature and relative proportions of the pyrolysis products obtained from these two systems. To obtain this equivalence, however, it was necessary to operate the filament at a temperature 70 °C below the Curie point.

Alexeeva and Khramova[112] compared a filament unit, a quartz tube furnace, and a Curie-point pyrolyser for the thermal breakdown of polyisoprenes and polybutadienes with a range of differing microstructures, styrene–butadiene rubbers, and blends of homopolymers. A temperature range of 420—980 °C was used. It was found that the furnace unit gave its most characteristic patterns at the lowest temperature (420 °C), but with only partial decomposition and without differentiating polymers of contrasting microstructure. Differences in microstructure were observed with the Curie-point pyrolyser (a commercial unit) but it is not clear if this is also true of the filament pyrolyser. Generally, the filament (a laboratory-made unit) and Curie-point units gave better results in terms of discrimination than the furnace type. The sample sizes are unfortunately not quoted.

In the specialised field of non-volatile petroleum fractions of similar composition, Leplat[35] found that a furnace operated with a static atmosphere gave better differentiation of crude oils, oil-bearing rock, *etc.* than did a filament pyrolyser operating in a through flow of carrier gas. Perhaps slight differences in the self-catalytic effects of the samples led to different secondary reactions.

An investigation of the activity of g.c. support materials was carried out in part by comparing their performance as packings in a stainless-steel pyrolysis furnace.[90] Other materials (including steel, glass and quartz wools, and gold powder) were also used. The amount of pyrolytic breakdown at temperatures of 400, 450, and 500 °C increased according to the packing material present; gold gave least, then quartz, glass, and the commercial g.c. packings in that order. It therefore appears that gold causes least catalytic activity, but the surface area of the packings is unknown, and unless it is similar for the various materials this reduces the usefulness of the results, though Fanter et al.[113] also concluded that an open gold tube was preferable to quartz due to its lesser catalytic activity, opposing the view of Goforth and Harris.[95]

Of particular interest are comparisons of results from different laboratories, which showed that, despite wide variation in pyrolyser types and conditions, similar results can be obtained (at least with some samples), including qualitative analysis of C_6—C_8 hydrocarbons[114] and quantitation of styrene monomer from a homopolymer.[115]

A wider view is taken when pyrolysis is compared with other fragmentation techniques, particularly mass spectrometry. The possible similarities of these two techniques have been referred to from time to time. Maccoll[116] gave several examples of analogous processes taking place under the influence of heat, light, and electron impact. These occur despite the range of energy states involved in these three methods of excitation, from vibrationally excited molecules through electronically excited molecules, to ions. In p.g.c., however, with pressures at the pyrolyser normally in excess of an atmosphere, excited species are usually in higher concentration than in the low-pressure environment of the mass spectrometer. Similarities between pyrolytic breakdown in p.g.c. and electron-impact decomposition in m.s. have been observed,[59] but the relatively high pressures used with p.g.c. reduce the likelihood of unimolecular breakdown, which is the normal mode of breakdown in m.s. Free-radical chain mechanisms frequently account for the breakdown in p.g.c.,[118] and, as pointed out by Kelly and Wolf,[117] this pyrolytic breakdown is unlikely to lead to fragments resembling those in the electron-impact breakdown of m.s. Since the extent of free-radical mechanisms on pyrolysis is difficult to predict, and can vary with sample size and conditions, existing m.s. data are unlikely to be of much use in interpreting pyrograms.[108]

References for Chapter 2

1   Grime, D., *The British Ink Maker*, 1967, 222.
2   Harms, D. L., *Analyt. Chem.*, 1953, **25**, 1140.
3   Mackillop, D. A., *Analyt. Chem.*, 1968, **40**, 687.
4   Coleman, W. E., Scheel, L. D., Kupel, R. E., and Larkin, R. L., *Amer. Ind. Hygiene Assoc. J.*, 1968, **29**, 33.
5   Davidson, W. H. T., Stanley, S., and Wragg, A. L., *Chem. and Ind.*, 1954, 1356.
6   Bigley, D. B., and May, R. W., *J. Chem. Soc. (B)*, 1967, 557.
7   Davis, A., and Golden, J. H., *European Polymer J.*, 1968, **4**, 581.
8   Tsuchiya, Y., and Sumi, K., *J. Polymer. Sci., Part A*, 1969, **7**, 3151.
9   Doue, F., and Guiochon, G., *J. Phys. Chem.*, 1969, **73**, 2804.
10 Barlow, A., Lehrle, R. S., Robb, J. C., and Sunderland, D., *Polymer*, 1967, **8**, 523.

11   Meier, J., Akermann, F., and Guenthard, Hs., *Helv. Chim. Acta*, 1968, **51**, 1686.
12   Sutton, R., and Harris, W. E., *Canad. J. Chem.*, 1968, **46**, 2623.
13   Levy, R. L., and Fanter, D. L., *Analyt. Chem.*, 1969, **41**, 1465.
14   Brauer, G. M., in 'Techniques and Methods of Polymer Evaluation,' Vol. II, Marcel Dekker Inc., New York, 1970, p. 41.
15   Jones, C. E. R., and Moyles, A. F., *Nature*, 1961, **191**, 663.
16   Levy, R. L., *J. Gas Chromatog.*, 1967, **5**, 107.
17   Farré-Ruis, F., and Guiochon, G., *Analyt. Chem.*, 1968, **40**, 998.
18   (a) Levy, R. L., *Chromatog. Rev.*, 1966, **8**, 48; (b) Walker, J. Q., *Chromatographia*, 1972, **5**, 547.
19   Jones, C. E. R., and Reynolds, G. E. J., *Reports Progr. Appl. Chem.*, 1969, **54**, 518.
20   Perry, S. G., *Adv. Chromatog.*, 1969, **7**, 221.
21   McKinney, R. W., in 'Ancillary Techniques of Gas Chromatography', ed. Ettre, L. S., and McFadden, W. H., Wiley–Interscience, London and New York, 1969, p. 55.
22   Coupe, N. B., Jones, C. E. R., and Perry, S. G., *J. Chromatog.*, 1970, **47**, 291.
23   Burrows, H. J., and Callam, D. H., *J. Chromatog.*, 1970, **53**, 566.
24   Romováček, J., and Kubát, J., *Analyt. Chem.*, 1968, **40**, 1119.
25   Liddicoet, T. H., and Smithson, L. H., *J. Amer. Oil Chemists' Soc.*, 1965, **42**, 1098.
26   Hewitt, G. C., and Witham, B. T., *Analyst*, 1961, **86**, 643.
27   Ettre, K., and Varaldi, P. F., *Analyt. Chem.*, 1963, **35**, 69.
28   Armitage, F., *J. Chromatog. Sci.*, 1971, **9**, 245.
29   Levy, R. L., and Wolf, C. J., *J. Chromatog. Sci.*, 1970, **8**, 524.
30   Levy, R. L., Wolf, C. J., Grayson, M. A., Gilbert, J., Gelpi, E., Updegrove, W. S., Zlatkis, A., and Oró, J., *Nature*, 1970, **227**, 148.
31   Oró, J., Updegrove, W., McReynolds, J., Ibanez, J., Gil-Av, E., Flory, D., and Zlatkis, A., *J. Chromatog. Sci.*, 1970, **8**, 297.
32   Deur-Šiftar, D., Bistricki, T., and Tandi, T., *J. Chromatog.*, 1966, **24**, 404.
33   Prosser, R. A., Stapler, J. T., and Yelland, W. E. C., *Analyt. Chem.*, 1967, **39**, 694.
34   Tsuge, S., Okumoto, T., and Takeuchi, T., *Bull. Chem. Soc. Japan*, 1969, **42**, 2870.
35   Leplat, P., *J. Gas Chromatog.*, 1967, **5**, 128.
36   Porter, R. S., Hoffman, A. S., and Johnson, J. F., *Analyt. Chem.*, 1962, **34**, 1179.
37   Shono, T., and Shinra, K., *Analyt. Chim. Acta*, 1971, **56**, 303.
38   O'Mara, M. M., *J. Polymer Sci.*, *Part A-1*, 1971, **9**, 1387.
39   Oró, J., Updegrove, W. S., Gibert, J., McReynolds, J., Gil-av, E., Ibanez, J., Zlatkis, A., Flory, D. A., Levy, R. L., and Wolf, C. J., *Science*, 1970, **167**, 765.
40   Perry, S. G., *J. Chromatog. Sci.*, 1969, **7**, 193.
41   Coupe, N. B., Jones, C. E. R., and Perry, S. G., *J. Chromatog.*, 1970, **47**, 291.
42   Jennings, E. C., and Dimick, K. P., *Analyt. Chem.*, 1962, **34**, 1543.
43   Barlow, A., Lehrle, R. S., and Robb, J. C., *Polymer*, 1961, **2**, 27.
44   Stanford, F. G., *Analyst*, 1965, **90**, 266.
45   Dimbat, M., and Eggertsen, F. T., *Microchem. J.*, 1965, **9**, 500.
46   Lehrle, R. S., and Robb, J. C., *J. Gas Chromatog.*, 1967, **5**, 89.
47   Franc, J., and Blaha, J., *J. Chromatog.*, 1964, **14**, 340.
48   Nelson, D. F., and Kirk, P. L., *Analyt. Chem.*, 1962, **34**, 899.
49   Perry, S. G., *J. Gas Chromatog.*, 1964, **2**, 93.
50   Lehmann, F. A., and Brauer, G. M., *Analyt. Chem.*, 1961, **33**, 673.
51   Zulaica, J., and Guiochon, G., *Analyt. Chem.*, 1963, **35**, 1724.
52   Krejci, M., and Deml, M., *Coll. Czech. Chem. Comm.*, 1965, **30**, 3071.
53   Levy, R. L., Fanter, D. L., and Wolf, C. J., *Analyt. Chem.*, 1972, **44**, 38.
54   Cogliano, J. G., *Rev. Sci. Instr.*, 1963, **34**, 439.
55   Szymanski, A., Salinas, C., and Kwitowski, P., *Nature*, 1960, **188**, 403.
56   Andrew, T. D., Phillips, C. S. G., and Semlyen, J. A., *J. Gas Chromatog.*, 1963, **1**, 27.
57   Janák, J., in 'Gas Chromatography 1960,' ed. Scott, R. P. W., Butterworths London, 1960, p. 387.
58   Giacobbo, H., and Simon, W., *Pharm. Acta Helv.*, 1964, **39**, 162.

59 Simon, W., Kriember, P., Voellmin, J. A., and Steiner, A., *J. Gas Chromatog.*, 1963, **1**, 53.
60 Bühler, C. L., and Simon, W., *J. Chromatog. Sci.*, 1970, **8**, 323.
61 Thompson, W. C., *Lab. Practice*, 1969, **18**, 1074.
62 Johns, T., and Morris, R. A., *Adv. Spectroscopy*, 1965, 361.
63 Sternberg, J. C., and Litle, R. L., *Analyt. Chem.*, 1966, **38**, 321.
64 Martin, S. B., *J. Chromatog.*, 1969, **2**, 272.
65 Martin, S. B., and Ramstad, R. W., *Analyt. Chem.*, 1961, **33**, 982.
66 Wiley, R. H., and Veeravagu, P., *J. Phys. Chem.*, 1968, **72**, 2417.
67 Folmer, O. F., and Azzaraga, L. V., *J. Chromatog. Sci.*, 1969, **7**, 665.
68 Guran, B. T., O'Brien, R. T., and Anderson, D. H., *Analyt. Chem.*, 1970, **42**, 115.
69 Butterworth, A., Home Office CRE Report No. 81, 1973.
70 Ristau, W. T., and Vanderborgh, N. E., *Analyt. Chem.*, 1970, **42**, 1848.
71 Ristau, W. T., and Vanderborgh, N. E., *Analyt. Chem.*, 1971, **43**, 702.
72 Ristau, W. T., and Vanderborgh, N. E., *Analyt. Chem.*, 1972, **44**, 359.
73 Kojina, J., and Morishita, F., *J. Chromatog. Sci.*, 1970, **8**, 471.
74 Folmer, O. F., *Analyt. Chem.*, 1971, **43**, 1057.
75 Fanter, D. L., Levy, R. L., and Wolf, C. J., *Analyt. Chem.*, 1972, **44**, 43.
76 Liehtmann, D., and Ready, J. F., *Phys. Rev. Letters*, 1963, **10**, 342.
77 Berkowitz, J., and Chupka, W. A., *J. Chem. Phys.*, 1964, **40**, 2735.
78 Biscar, J. P., *J. Chromatog.*, 1971, **56**, 348.
79 (a) Biscar, J. P., *Analyt. Chem.*, 1971, **43**, 982; (b) Merritt, C., Sacher, R. E., and Petersen, P. A., *J. Chromatog.*, 1974, **99**, 301.
80 Juvet, R. S., Smith, J. L. S., and Li, K.-P., *Analyt. Chem.*, 1972, **44**, 49.
81 Simmonds, P. G., Shulmah, G. P., and Stembridge, C. H., *J. Chromatog. Sci.*, 1969, **7**, 36.
82 Robb, E. W., and Westbrook, J. J., *Analyt. Chem.*, 1963, **35**, 1645.
83 Keulemans, A. I. M., and Perry, S. G., in 'Gas Chromatography 1962,' ed. van Swaay, M., Butterworths, London, 1962, p. 356.
84 Wolf, T., and Rosie, D. M., *Analyt. Chem.*, 1967, **39**, 725.
85 Cramers, C. A. M., and Keulemans, A. I. M., *J. Gas Chromatog.*, 1967, **5**, 58.
86 Denkar, W. D., and Wolf, C. J., *J. Chromatog. Sci.*, 1970, **8**, 534.
87 Groenendyk, H., Levy, E. J., and Sarner, S. F., *J. Chromatog. Sci.*, 1970, **8**, 115.
88 Walker, J. Q., and Maynard, J. B., *Analyt. Chem.*, 1971, **43**, 1548.
89 Abbot, S. D., Hall, R. C., and Giam, C. S., *J. Chromatog.*, 1969, **45**, 317.
90 Verzele, M., Van Cauwenberghe, K., and Bouche, J., *J. Gas Chromatog.*, 1967, **5**, 114.
91 Gough, T. A., and Walker, E. A., *J. Chromatog. Sci.*, 1970, **8**, 134.
92 Dhont, J. H., *Analyst*, 1964, **89**, 71.
93 Tinkelenberg, A., *J. Chromatog. Sci.*, 1970, **8**, 721.
94 Sternberg, B. C., Krull, I. H., and Friedel, G. D., *Analyt. Chem.*, 1966, **38**, 1639.
95 Goforth, R. R., and Harris, W. E., in 'Gas Chromatography 1968,' ed. Harbourn, C. L. A., and Stock, R., Institute of Petroleum, London, 1969, p. 261.
96 Bower, K., Littlewood, A. B., and Phillips, C. S. G., *J. Inorg. Nuclear Chem.*, 1960, **15**, 316.
97 Levy, E. J., and Paul, D. J., *J. Gas Chromatog.*, 1967, **5**, 136.
98 Levy, E. J., U. S. P. 3 425 807 (1969).
99 Groenendyk, H., Levy, E. J., and Sarner, S. F., *J. Chromatog. Sci.*, 1970, **8**, 599.
100 Walker, J. Q., and Wolf, C. J., *Chem. Eng. News*, 1967, Sept. 25, p. 47.
101 Walker, J. Q., and Wolf, C. J., *Analyt. Chem.*, 1968, **40**, 711.
102 Walker, J. Q., and Wolf, C. J., in 'Gas Chromatography 1968,' ed. Harbourn, C. L. A., and Stock, R., Institute of Petroleum, London, 1969, p. 385.
103 Merritt, C., and DiPietro, C., *Analyt. Chem.*, 1972, **44**, 57.
104 Smalldon, K. W., personal communication.
105 Jackson, M. T., and Walker, J. Q., *Analyt. Chem.*, 1971, **43**, 74.
106 Tukeda, I., and Mushiko, Y., *Kogyo Kagaku Zasshi*, 1968, **71**, 1627.

107 Walker, J. Q., and Wolf, C. J., *J. Chromatog. Sci.*, 1970, **8**, 513.
108 Kossiakoff, A., and Rice, F. O., *J. Amer. Chem. Soc.*, 1943, **65**, 590.
109 Jones, C. E. R., and Reynolds, G. E. J., *J. Gas Chromatog.*, 1967, **5**, 25.
110 Van Cauwenberghe, K., Vandewalle, M., and Verzele, M., *J. Chromatog. Sci.*, 1969, **7**, 698.
111 Galvin-Vacherot, M., *European Polymer J.*, 1971, **7**, 1445.
112 Alekseeva, K., and Khramova, L. P., *J. Chromatog.*, 1972, **69**, 65.
113 Fanter, D. L., Grayson, M. A., and Wolf, C. J., 154th National American Chem. Soc. Meeting, Chicago, 1967, No. B60.
114 Fanter, D. L., Walker, J. Q., and Wolf, C. J., *Analyt. Chem.*, 1968, **40**, 2168.
115 Jones, C. E. R., and Reynolds, G. E. J., *Brit. Polymer. J.*, 1969, **1**, 197.
116 Maccoll, A., 'Modern Aspects of Mass Spectrometry,' Proceedings of 2nd Conference of the NATO Advanced Study Institute for Mass Spectrometry, 1966, p. 143.
117 Kelly, J. D., and Wolf, C. J., *J. Chromatog. Sci.*, 1970, **8**, 583.
118 Lille, U., and Kundle, H., *J. Chromatog.*, 1972, **69**, 59.
119 Nelson, K. H., Groeper, L. H., and Lysyi, I., *Chem. Instrum.*, 1970, **2**, 355.

3 Applications of Pyrolysis–Gas Chromatography

It has been said that p.g.c. is applicable to any material that is capable of being thermally decomposed to yield a pyrolysate of which some portion can be separated by g.c.[1] The value of the results will clearly vary with the nature of the material, however.

Recent reviews have considered the use of p.g.c. for identification of synthetic polymers and for determination of polymer microstructure,[2-5] and for investigation of biological macromolecules;[6] and there have been some more general reviews;[1,7,8] there has also been a recent general bibliography of the uses of the technique.[9]

The reviews on specified types of sample reflect the more important areas in which p.g.c. is used. Further to these applications, p.g.c. is employed in the identification of a variety of biochemical materials and organisms, and of both involatile and volatile non-polymeric organic compounds. The technique can be applied as either a qualitative or quantitative tool for analysis and is also used under carefully controlled conditions in kinetic experiments for the elucidation of reaction mechanisms. Of the above applications, the analysis of synthetic polymers is probably the most common, and will be considered first.

Synthetic Polymers

This section on p.g.c. of synthetic polymers is intended to give an indication of the scope of work published to date, with examples taken mainly from the most recent work. The reviews[1-9] cited above give excellent coverage of earlier work. The subject is sometimes divided into work concerned with (a) the identification and differentiation of polymers in a qualitative way, often referred to as 'fingerprint pyrolysis', and (b) quantitative determinations of copolymers and polymer mixtures, which in the more sophisticated forms lead to the elucidation of polymer microstructure. This division is not always clear, since a 'fingerprint pyrogram', in the sense of a pyrogram unique to a particular sample, can only be obtained in some cases when the same detailed attention is given to the work as is necessary for determination of microstructure. In this chapter, those studies involving the identification of a range of polymer types, which may include some quantitation of pyrogram peaks, will comprise the section on fingerprint pyrolyses.

'Fingerprint' Pyrograms

Although the range of polymers that have been subjected to p.g.c. is large, as witnessed in the great number of papers on the subject, most deal with a limited range of polymer types. Only a few detail the pyrolysis of wide ranges of polymers. Nelson, Yee, and Kirk[10] pyrolysed a series of commercial plastics, using 0.2—0.5 mg samples, on a filament pyrolyser at 'dark red heat', for 10 s;

53

a packed column containing 5% silicone oil as liquid phase was used in the G.C. Where pyrograms were indistinguishable, pyrolyses at different temperatures were used to differentiate the samples. Cox and Ellis[11] pyrolysed 39 polymers representing a range of elastomers, plastics, textiles, and surface coatings. Their pyrolyser was of the tubular quartz furnace type and packed with quartz wool; it was operated at 700 °C. Their column contained 16.7% silicone oil as liquid phase, and the data were reduced to the form of a block diagram (Figure 3.1), retention times being related to the last significant peak

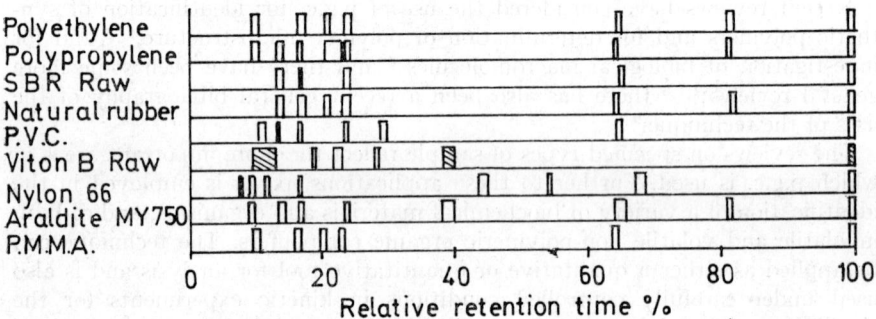

Fig. 3.1   Chart of reduced retention times for some polymers. Negative peaks are indicated by cross-hatching. Only peaks which have well-defined relative retention times have been included, but, if the sample is small or the instrument sensitivity low, smaller peaks may not be shown in the pyrogram
(Reproduced, with permission, from *Analyt. Chem.*, 1964, **36**, p. 95)

of retention time less than 15 minutes. A block diagram is a simple and frequently used way of transmitting such data, but the danger of rendering the reduced pyrograms difficult to recognise when overlapping peaks are present is seen by comparing the pyrogram of polypropylene (Figure 3.2) with the reduced data. Groten[12] pyrolysed over 150 materials, including plastics, elastomers, resins and natural products, homopolymers, blends, and copolymers. The pyrolyser was a platinum filament and the samples (about 1 mg) were held in a quartz capillary in the centre of the tube. Pyrolysis temperature was a rather high 950 °C, held for 26 seconds. The stationary phase was 20% polyglycol or 5% polyglycol, the latter being used for the cellulose esters. Where ambiguous pyrograms were obtained, pyrolysis time and/or temperature were adjusted to establish identity. Quantitative determinations were carried out on some of the polymers. Some differences were illustrated by line diagrams representing the pyrograms, using a log–log plot of peak height *versus* retention time. For general use, such diagrams could be confusing, as is the case for those of Cox and Ellis.[11] Berton[13] used line diagrams similar to the usual presentation of a mass spectrum to illustrate the pyrolyses of a range of polymers, including paints and textile fibres (the pyrolysis was conducted separately from the G.C.). The stationary phase was butyl phthalate.

P.g.c. has been used to 'fingerprint' groups of polymers of similar application but of differing chemical type. Several fingerprint methods for paint analysis have been described.[14-16,17a-c] P.g.c. is advantageous in the identification of polymers in paint in that it is insensitive to pigments which are in high concentration in paints; the fact that only a small sample is necessary is useful in forensic examinations[17d] and the investigation of works of art.

A field resembling paint, in that it employs a range of polymeric materials and is of forensic interest, is that of textile fibres. Haase and Rau[18] pyrolysed both natural and synthetic fibres to produce 'fingerprint' pyrograms. A similar group of fibres were pyrolysed by Janiak and Damerau;[19a] the published pyrograms in these studies were produced on different stationary phases, and different 'fingerprints' were produced for similar fibres. Bortniak et al.[19b]

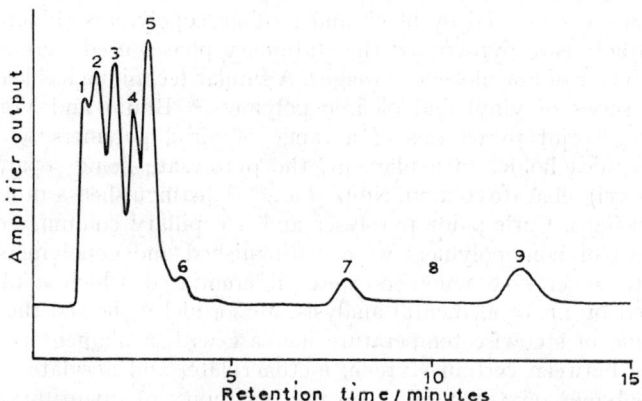

Fig. 3.2 Chromatogram for polypropylene. Peak 8 is not recorded because of the small sample and low instrument sensitivity used for the run
(Reproduced, with permission, from *Analyt. Chem.*, 1964, **36**, p. 93)

restricted their work to acrylic and modacrylic fibres, placing 27 acrylics and 14 modacrylics into 12 and 3 groups, respectively, the groups being defined largely by ratios of peak heights.

Other materials from which fingerprint pyrograms have been produced include rubber materials,[20] brake linings[21a] (the pyrolysable materials are organic resins in an asbestos matrix), adhesives,[21b] soil, [21c-e] hair,[21f] flame-retardant finishes,[21g] and 'ashless' additives of mineral oils separated from the oils by molecular distillation.[22] Nematollahi et al.[23] differentiated a number of plastics of medical interest by stepped-temperature pyrolysis on a filament, using no separating column, the pyrolysate being fed directly to a F.I.D. This method is on the periphery of our subject, as is the combination of thermogravimetry with g.c., which gives both pyrograms and thermogravimetry patterns from a few milligrams of sample.[24]

Work reported on limited ranges of polymers is usually concerned with quantitative analysis, though the distinction between such studies and 'finger-printing' is frequently vague and unimportant. The rest of this section on applications to synthetic polymers will be arranged under headings according to polymer type.

Vinyl Polymers

Barlow *et al.*[25] examined a range of vinyl-type polymers, using a filament operated at stepped temperatures of 150—950 °C to characterise the polymers; they used a single temperature (580 °C) for a quantitative analysis of vinyl copolymers and obtained values accurate to within one per cent. Differences between copolymers and mixtures of homopolymers were seen using the stepped-temperature method (Figure 3.3 h, i). Also, differences in polymer microstructure exemplified by block and random copolymers (Figure 3.3 j, k) yielded characteristic pyrograms; the stationary phases used were silicone oil and polyethylene of low molecular weight. A similar technique has been applied to another series of vinyl and olefinic polymers.[26] Braun and Vorendohre[27] obtained fingerprint pyrograms of a range of vinyl polymers pyrolysed at 800 °C on a mica holder in a filament, the pyrolysate being separated on a di(methylhexyl) sebacate column. Noffz *et al.*,[28,29] distinguished a range of vinyl polymers, using a Curie-point pyrolyser and a capillary column. Copolymers and mixtures of homopolymers were distinguished and copolymers of vinyl acetate with several comonomers were differentiated which could not be distinguished by i.r. or elemental analysis. McCormick[30] showed the ability of the technique of stepwise temperature increase with a filament pyrolyser to differentiate between certain styrene, methacrylate, and acrylate copolymers and homopolymer mixtures, and also the possibility of quantitation of such copolymers of known microstructure at fixed pyrolysis temperatures.

Most of the recent work on vinyl polymers has been concerned with quantitative studies, mainly with a view to elucidating microstructure. Jones and Reynolds[3] presented a critical review of this subject and showed the usefulness of the technique in differentiating random and block copolymers of butadiene–styrene. A second method, to establish total monomer polymerised, is necessary, and these results must then be compared with those from p.g.c. A random copolymer will yield less of the total styrene than a block copolymer with the same proportion of monomer. The variation of yield of styrene monomer from polymers containing different comonomers with a range of reactivity ratios with styrene was also investigated. The proportion of blocks of monomer units in block copolymers of styrene has been investigated in a similar study, based on a determination of copolymer composition derived by differential refracto-metry.[31] Another study illustrated the use of p.g.c. in the determination of the microstructure of styrene and other vinyl aromatic hydrocarbons copoly-merised with aliphatic dienes.[32] When a series of polystyrenes of decreasing molecular weight are pyrolysed, the monomer yield drops, the polymer being more stable at low molecular weights (although the stability of the polymer

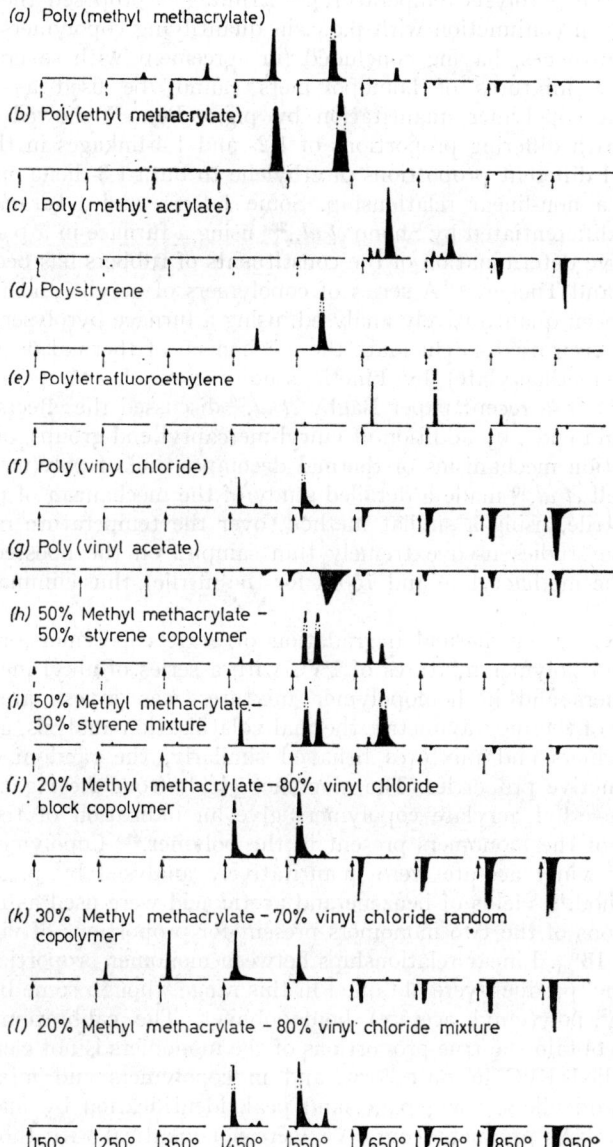

(a) Poly(methyl methacrylate)

(b) Poly(ethyl methacrylate)

(c) Poly(methyl acrylate)

(d) Polystryrene

(e) Polytetrafluoroethylene

(f) Poly(vinyl chloride)

(g) Poly(vinyl acetate)

(h) 50% Methyl methacrylate –
50% styrene copolymer

(i) 50% Methyl methacrylate–
50% styrene mixture

(j) 20% Methyl methacrylate–80% vinyl chloride
block copolymer

(k) 30% Methyl methacrylate–70% vinyl chloride random
copolymer

(l) 20% Methyl methacrylate–80% vinyl chloride mixture

150° 250° 350° 450° 550° 650° 750° 850° 950°

Fig. 3.3 Typical pyrograms obtained by the filament technique for the degradation of polymer samples
(Reproduced, with permission, from *Polymer*, 1961, **2**, p. 39)

depends on the pyrolysis temperature).[33] Armitage[34] proposed the use of i.r. spectrometry in conjunction with p.g.c. in quantifying copolymers of styrene and aliphatic dienes, having concluded (in agreement with several previous workers) that mixtures of homopolymers cannot be used as calibration standards for copolymer quantitation by p.g.c. Perry[35] showed that polybutadienes with differing proportions of 1,2- and 1,4-linkages in the polymer chain yielded different proportions of ethylene to buta-1,3-diene on pyrolysis, there being a non-linear relationship. Some 1,4-*cis*- and 1,4-*trans*-polybutadienes were differentiated by Shono *et al.*,[36a] using a furnace in a p.g.c. system. A quantitative determination of the constituents of rubbers has been reported by Krishen and Tucker.[36b] A series of copolymers of acrylic and methacrylic esters have been quantitatively analysed, using a furnace pyrolyser.[37]

P.g.c. has been used to elucidate the mechanism of thermal degradation of poly(methyl methacrylate) by kinetic studies over the temperature range 340—460 °C.[38] In a recent paper Bagby *et al.*[39] discussed the effects of modification of chain ends, by addition of lauryl-mercaptyl end-groups, on initiation and termination mechanisms of thermal decomposition of poly(methyl methacrylate). Bell *et al.*[40] made a detailed study of the mechanism of pyrolysis of polyacrylonitrile, using a similar method, over the temperature range 200—850 °C. These studies used extremely thin samples on the boosted filaments (200 Å for the methacrylate and 750 Å for the nitrile), thus eliminating thickness effects.

The behaviour on thermal degradation of graft copolymers, random copolymers, and polymer mixtures of PVC with a series of alkyl methacrylates as comonomers and in homopolymer mixtures was investigated, using a combination of thermogravimetry, thermal volatilisation analysis, and p.g.c.[41a] Graft copolymers and mixtures behaved similarly, the random copolymers having distinctive properties. The monomer yields from the p.g.c. of methyl methacrylate–ethyl acrylate copolymers give an indication of the sequence distribution of the monomers present in the polymer.[41b] Copolymers of vinyl chloride and vinyl acetate were quantitatively analysed by p.g.c., using a furnace method.[42] Yields of benzene and acetic acid were used as measures of the proportions of the two monomers present for proportions of vinyl acetate up to about 16%. Linear relationships between monomer proportions and the 'characteristic' product were obtained in this range, though some benzene was produced by poly(vinyl acetate) homopolymer. The calibration technique employed to obtain the true proportions of the monomers is not clearly stated. O'Mara studied PVC in pure form and in copolymers and mixtures with plasticisers and fillers, using p.g.c. and peak identification by mass spectrometry.[43] Pyrolysis products from PVC were not identical with those of Noffz *et al.*,[29] but not only were the pyrolysis conditions different (furnace as opposed to Curie-point, by Noffz) but the g.c. conditions differed sufficiently to make comparison difficult. PVC samples incorporating a series of phthalate plasticisers were pyrolysed and shown to give 'fingerprint' pyrograms consisting of breakdown products of PVC and the plasticisers without interaction of the two components. HCl is produced in stoicheiometric yields from unfilled PVC, but certain basic inorganic fillers absorb a proportion of the HCl. Random and alternating copolymers of vinyl chloride and methyl methacrylate have been examined using a Curie-point pyrolyser.[44]

It is of interest that, in the above studies of vinyl polymers, all three common types of pyrolyser (furnace, hot filament, and Curie-point) have been used at temperatures covering the range 430—700 °C, though the boosted filament type was particularly favoured for kinetic studies. Where hydrocarbons were the major pyrolysis products the G.C. columns most frequently utilised contained silicone oils and gums and Apiezon grease as the stationary phase, though other materials were used on occasion, particularly to study the most volatile products. Where more polar products were obtained, a much wider range of stationary phases were applied, including Porapak Q, polyglycol, diethylene glycol succinate, $\beta\beta$-oxydipropionitrile, and tricresyl phosphate.

Polyolefins
Recent work on polyolefins includes the 'fingerprinting' of a range of olefin polymers, ethylene homopolymer, ethylene–butene, ethylene–propylene, and ethylene–hex-1-ene copolymers, polyisobutene, propylene homopolymer, and a polypropylene copolymer, using a Curie-point pyrolyser, and based on pyrograms consisting of hydrocarbons up to C_7.[45] Although these polymers were differentiated at each of several pyrolysis temperatures in the range 480—770 °C, three high-density homopolymers of ethylene were only differentiated at the lower temperature (Figure 3.4). Low-density homopolymers were also differentiated, both from each other and from the high-density polymers; once again, the differences were most marked at 480 °C. The paper concludes with a discussion on the relationship of the pyrolysis product ratios with polymer microstructure. A furnace pyrolyser was used in the investigation of polypropylene tacticity by p.g.c.[46] It was found that the ratio of the iso-C_4:n-C_4 peaks could be used to distinguish isotactic polypropylene from the syndiotactic and atactic forms. A linear relationship between the (iso-C_4—n-C_4):iso-C_4 ratio and the proportion of isotactic polypropylene in a sample of polypropylene was found, enabling the latter proportion to be determined after constructing a simple calibration curve. Separation of diastereoisomers in the C_{15} region has also been used to differentiate polypropylenes of different tacticity.[29]

Galvin-Vacherot[47,48] used a filament pyrolyser in the investigation of polyisoprenes containing the three types of isoprene polymer-addition unit [1,4-(*cis* and *trans*), 3,4-, and 1,2-]. On pyrolysis, the 1,4- type were characterised by the production of dipentene and the 3,4-type by formation of 3,4-dimethyl-4-vinylcyclohexene; 1,2-type units gave isoprene. The yields of these products depend on the microstructure of the polymer and hence can be used to evaluate this latter property.

P.g.c. has been used for some time to 'fingerprint' commercial rubbers as found in vehicle tyres.[49] Quantitation of component rubber materials has been achieved.[50] Ternary blends of natural rubber, styrene–butadiene rubber, and ethylene–propylene–terpolymer in compounded, cured stocks have been quantitatively determined, using calibration curves based on selected pyrogram peaks.[51] The peaks chosen in this latter study were affected by other rubbers, such as neoprene, polybutadiene, chlorobutyl, and butyl. Only the most volatile products were eluted, but quantitation was successful despite the relatively complex breakdown of ethylene–propylene-terpolymer, which, when mixed with the other ingredients, led to a pyrogram containing 17 peaks.

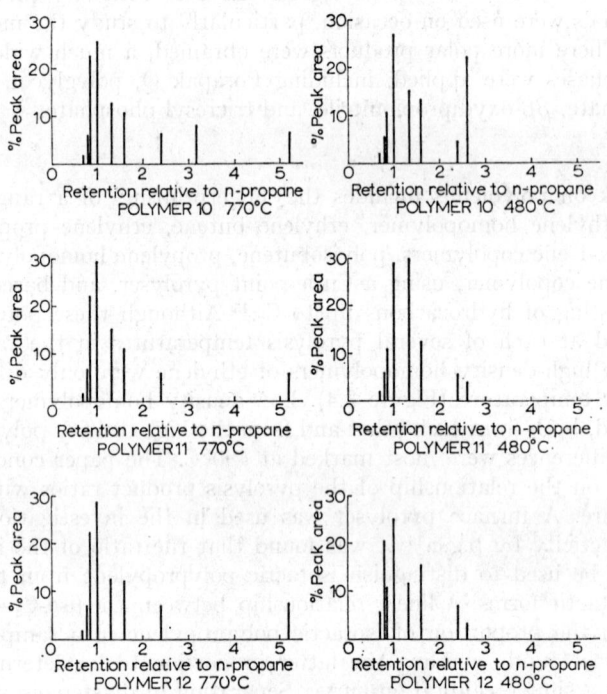

Fig. 3.4   Pyrograms of high-density ethylene homopolymers at 770 and 480 °C
(Reproduced, with permission, from *J. Chromatog. Sci.*, 1969, **7**, p. 106)

In the quoted studies on polyolefins and rubbers, all three common types of pyrolyser were used over the temperature range 350—700 °C and, despite the restricted nature of the pyrolysis products, a range of stationary phases were used. Phases included polyglycols, polypropylene carbonate, silicone gum, Apiezon L, and tricresyl phosphate.

Polyurethanes

Urethane foams have been identified[52a] and their constituent isocyanates quantitatively determined.[52b] A series of commercial polyurethane elastomers were pyrolysed and a scheme for identifying the 'fingerprint' pyrograms was presented.[53]

Polycarbonates

Tsuge *et al.*[54] differentiated polycarbonates based on diphenylolpropane (the most common types) according to the method of synthesis (melt or solvent method). The two methods produce different end-groups on the chains, hence the differentiation by p.g.c. Since the pyrogram differences are caused by end-groups, quantitation of the variable peaks can be used to estimate molecular weight.

Chlorinated Polymers

The microstructure of vinylidene chloride–vinyl chloride copolymers was investigated by Tsuge *et al.*[55] The yields of benzene and mono-, di-, and tri-chlorobenzene and vinylidene chloride were related to the polymer moieties which were most likely to have produced them, and hence the proportions of these moieties present in the polymer were estimated. These results were shown to agree with proportions calculated by copolymerisation theory. The chlorine content and microstructure of chlorinated polyethylenes, chlorinated polybutadienes, chlorinated PVCs, and chlorinated polystyrenes were estimated from degradation products (excluding hydrogen chloride), using a knowledge of the mechanisms of thermal degradation,[56,57] as in the previous study.

Polyamides

The quantitation of Nylon 6-66 copolymers and mixtures was carried out by measuring the relative yield between cyclopentanone and ϵ-caprolactam on pyrolysis. Water adsorbed on the polymer was found to reduce the precision of the method,[58] possibly due to the introduction of a variable proportion of ionic decomposition, as has been shown to be of importance in the breakdown of some polyesters.[59]

Not only organic polymers yield useful information by p.g.c.; any thermally degradable polymer yielding sufficiently volatile fragments which can be detected will give a pyrogram. For example, the mechanism of thermal decomposition of branched-chain methylsiloxane polymers has been investigated, using a normal G.C. set-up with a flame ionisation detector.[60]

Much of the qualitative work reviewed here and in previous works has been performed with such widely varying conditions of pyrolysis and gas chromatography as to make comparison impossible. It is perhaps inevitable that, when two closely related polymers are to be differentiated, only very specific conditions will perform the task; however, it is to be hoped that the future will see greater emphasis on standardisation of conditions, both of pyrolysis and of gas chromatography. A similar situation prevails in quantitative studies where, despite the publication of work (largely on depolymerisation mechanisms[61],[62]) showing the necessity of minimising secondary reactions by the use of microgram sample sizes, *etc.*, much recent work quoted in this chapter has involved the use of milligram sample sizes and/or furnace pyrolysers, with the attendant dangers of secondary reactions inherent in such practice (Chapter 2).

Involatile Non-polymeric Organic Materials

Although it is probably true that most p.g.c. applications have been in the field of polymers (mainly synthetics), useful information can be obtained by using p.g.c. on other involatile materials. Plasticisers, which are commonly involatile esters, are important additives to such polymers as PVC. Hagen[63] pyrolysed ten plasticisers on a filament at 580—730 °C. Using three different stationary phases to separate the products, all ten were differentiated. Zulaica and Guiochon[64] differentiated seven plasticisers injected into a G.C. (without pyrolysis), using two stationary phases, and employed Kovats indices in identifying the ester plasticisers. Quantitative analyses were then performed on mixtures of plasticisers present in PVC, using a filament pyrolyser to liberate the plasticisers from the PVC matrix without pyrolysis of the plasticisers.

Surfactants, detergents, and soaps have been subjected to p.g.c. A furnace was used at 650 °C to pyrolyse a series of surfactants individually, both in mixed form and when present in commercial detergents.[65] In the latter two cases, quantitative results were obtained for some of the surfactants by p.g.c. Aryl sulphonic acids and their salts, dried from aqueous solutions, were pyrolysed in a furnace at 710—790 °C to obtain quantitative analyses by measuring the yields of sulphur dioxide and hydrocarbon.[66] The samples were rendered alkaline to eliminate interference from sulphur dioxide produced from sulphuric acid, and carbohydrazide was added to increase the yield of hydrocarbon without itself interfering with the pyrogram.

Long-chain fatty acids were pyrolysed in admixture with acetates in a furnace at 400—600 °C to yield alkyl methyl ketones.[67] These products are easily eluted from a G.C. and can therefore be used to detect and quantify the fatty acid content of commercial soaps, pyrolysed as their barium soaps. The method is compared with the use of methyl ester formation followed by g.c. A laser pyrolysis has been used to differentiate the sodium salts of C_6 n- and iso-caproates.[68]

Synthetic drugs have been identified by p.g.c.[69] Nelson and Kirk[70],[71] differentiated 27 barbiturates in therapeutic use, pyrolysing them on a filament at 'red heat' and separating the products on three different polyglycol phases. Many products were identified, and related to possible decomposition mechanisms.

Abbot *et al.*[72] decarboxylated the herbicide picrolam (4-amino-3,5,6-tri-chloropicolenic acid) in a home-made pyrolyser fitted to a G.C. with an electron-capture detector. The system is said to provide a pyrogram that is free from impurity peaks, thus allowing detection and quantitation of picrolam in trace quantities in organic matrices.

Pyrolysis of alkyl onium halides yields tertiary amines and methyl halides. P.g.c. of these compounds can be used for identification at the 100ng level.[73] Under the conditions used (*i.e.* filament pyrolysis at 500 °C), no further breakdown occurs, allowing quantitation of mixtures of these compounds.

An interesting application of p.g.c. is for the identification of organometallic compounds. Zinc, lead, and potassium dialkyl thionothiophosphates were characterised, using a quartz-filled injection port as pyrolyser in a p.g.c. system.[74] The zinc compounds are found as additives in lubricating oils. Perry[75] separated these zinc compounds from oils by t.l.c. and then lifted them from the t.l.c. medium with methylene chloride. The dried organometallics were pyrolysed on a 'filament' dish in microgram quantities; comparison with the previous work of Legate and Burnham[74] on the lead derivatives showed good agreement, despite the different pyrolysis conditions. Perry discussed possible mechanisms, and the potentialities and limitations of the technique, for organometallics and other applications.

P.g.c. is of great potential in 'fingerprinting' and in the elucidation of the composition of involatiles associated with fossil fuels. Coal and its derivatives have been differentiated, using an unusual furnace pyrolyser.[76] The kinetics of benzene and toluene production were also studied. Useful analytical information can be obtained quickly by p.g.c. on the otherwise relatively intractable bituminous materials. Leplat[77] preferred a static-atmosphere furnace at 600 °C to obtain maximum discrimination between asphalts and potentially oil-bearing rocks. Poxon and Wright[78] used a Curie-point pyrolyser to characterise seven bitumens. Two sets of pyrolyses, one at 610 °C and another at 980 °C, were required to obtain complete discrimination with a substrate coated with Apiezon L in the G.C. column. However, the differences were slight, and characterisation was dependent on the statistical interpretation of peak areas. The hydrocarbon content of oil shales was evaluated, using a laser pyrolyser that simultaneously released the hydrocarbon from the inorganic matrix and pyrolysed it,[79] to give mainly methane, ethylene, and acetylene; the yields of these gases are a reliable means of estimating total hydrocarbon content, assuming that each shot vaporises a constant quantity of shale.

A much smaller proportion (if any) of organic matter is expected to be present in rock of extraterrestrial origin as compared with oil-bearing strata. Such rock, in the form of meteorites,[80,81] lunar rock,[82,83] and earth sands and shales as models for Martian soils,[84] has been subjected to p.g.c. in furnace pyrolysers as a means of detecting potential life-supporting chemicals. In each case of extraterrestrial material, organic matter has been discovered; however, the traces found are too small to rule out terrestrial contamination as their source.

Volatile Compounds

The p.g.c. of volatile compounds is a field of increasing importance. For example, kinetic measurements, independent of volatilisation phenomena, can be made under precisely controlled conditions. Perhaps the most interesting application, however, is in the identification of g.c. effluents; the technique of g.c.–p.g.c. is discussed in Chapter 2.

Hydrocarbons

The group of compounds that have received the most attention in gas-phase pyrolysis are the hydrocarbons with carbon numbers up to about 19. Pyrolyses have been carried out in packed and open tubular furnaces of glass and metallic construction; in the main, heating has been applied externally, but an electric arc has been used.[85] Keulemans and Perry[86] pyrolysed a series of hydrocarbons up to C_{10}, mainly at 500 °C. The influence of sample size and temperature were discussed and the cracking patterns interpreted in terms of the parent molecules' structures.

Holman *et al.*[87] pyrolysed sixteen normal, branched, and cyclic aliphatic hydrocarbons up to C_{19} in a Pyrex tube at 600 °C, subjecting the pyrolysates to g.c.–m.s. The interpretation of the pyrolysis breakdown patterns was likened to that of mass spectral fragmentation patterns. Cramers and Keulemans[88] pyrolysed 100 alkanes and alkenes (C_5–C_{16}), obtaining different pyrograms (using a capillary column) for each, except for *cis/trans*-isomers, which differed only in percentage conversion. Product distributions as a function of residence time, pyrolysis temperature, and sample size were investigated, and the use of the technique in kinetic measurements was demonstrated. The data of Cramers and Keulemans have recently been interpreted[89] on the basis of the Rice free-radical theory.[90] Fanter *et al.*[91] pyrolysed a similar series of 83 hydrocarbons of carbon number 6—10 and confirmed that all but *cis/trans*-isomers differed, as with the work of Cramers and Keulemans.[88] The relative ease of differentiation of six C_6H_{12} isomers by p.g.c. compared with the use of m.s. was demonstrated, as was the similarity in pyrolysis patterns obtained by three different laboratories, using different furnace conditions. An arc pyrolyser was used and the carrier gas varied from helium to hydrogen to differentiate a series of hydrocarbons and compounds containing ether, alcohol, ketone, and halide groups.[85] Pyrolysis followed by catalytic hydrogenation has been used to identify a number of normal and branched hydrocarbons.[92] The degradation mechanisms were discussed. The relative quantities of branched- and straight-chain alkenes were found to differ with the position of branching and molecular weight in the parent compound, thus permitting the deduction of unknown branched structures. The degree of branching in alkylbenzenes has been determined in a similar way by measuring the ratios of styrene to α-methylstyrene[93a] and light pyrolytic fractions in the pyrolysate;[93b] the alkylbenzenes are obtained from their sulphonates, which are used in detergents. The biodegradability of these substances depends on the degree of branching; hence this p.g.c. technique is of importance in the monitoring and subsequent control of water pollution. The kinetic parameters of the thermal decomposition of methyl decanoate and hexadecane in a gold tube pyrolyser have been studied.[94] The mechanism as predicted by Rice theory agreed with the cracking pattern obtained, which was largely independent of sample size. The four isomers of trimethylpentane were characterised by p.g.c., using a gold tube pyrolyser at 680°C.[95] The products were shown to be in agreement with Rice free-radical theory and with two previous reports of gas-phase pyrolysis of these compounds, despite wide variation in conditions.

The good agreement in cracking patterns between different workers and with free-radical theory in the above reports shows that pyrolyser designs are available which give pyrolyses largely free from secondary reactions, thus enhancing the value of the technique on an interlaboratory scale and increasing the chances of identifying an unknown from its fragmentation pattern alone, without the need for comparison with standard pyrograms.

Identification of Functional Groups

Possibilities of using p.g.c. for the identification of functional groups as well as hydrocarbon 'backbone' structures were explored by Dhont,[96] and have recently received more attention. Alkanes, alkenes, alcohols, thiols, and methyl esters have been pyrolysed in a quartz tube to yield pyrograms characteristic of the parent compound types,[97] and been related to the parent compounds by Rice theory. Useful 'fingerprint' pyrograms of similar materials have been obtained with a gold tube pyrolyser[98] and an electric arc. These pyrograms are complex and difficult to interpret without knowledge of the parent compound. Some indication of the latter is obtained if the system is of the tandem g.c. type, *i.e.* g.c.–p.g.c., since a retention time for the parent will be obtained in the first G.C. and possibly in the second, if there is only partial breakdown in the pyrolyser and the second G.C. column can elute the parent. The proportion of breakdown is an important characteristic, as we have seen with *cis–trans*-isomers; it also helps in determining the type of functional groups present. A range of temperatures, giving between 10 and 90% breakdown, has been used in a similar way.[99] A more simple approach to the determination of functional groups is the quantitation of the 'small molecules' pyrogram. Small molecules from pyrolysis (CH_4, C_2H_4, CO, CO_2, H_2O, H_2S, HCl, NH_3, *etc.*) are separated and relative peak areas determined. The relative area may be related to the total area of all peaks,[100a] or merely paired to give ratios; such relative areas are found to be fairly constant for certain functional groups, exemplified by alcohols, aldehydes, ketones, ethers, esters, thiols, and sulphides.[100b] Knowledge thus obtained of the functional groups present, and of the backbone structure deduced from the hydrocarbon part of the pyrolysate, can be combined with retention-volume data obtained on a primary separation column to locate the unknown compound in a homologous series (Figure 3.5). A functional group analyser has been combined with catalytic reactors to enable C, H, N, O determinations to be made with the same unit.[101] Until all processes involved in the g.c.–p.g.c. system are standardised, it seems likely that a bank of data will have to be built up for each system, and as yet too few compounds have been investigated to prove the validity of the 'small molecule' pyrogram as a specific means of analysis for functional groups. At the present time the use of fingerprint pyrograms of limited ranges of compounds, such as straight- and branched-chain methyl esters[102] or aliphatic alcohols,[103] is still of most importance in analytical practice.

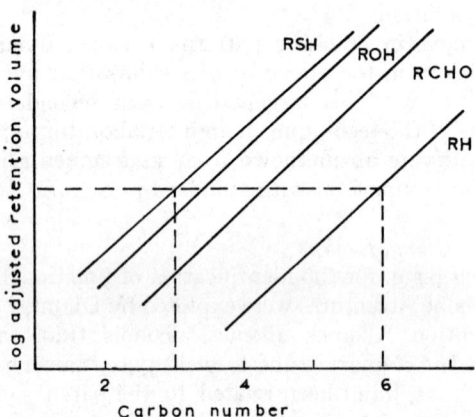

Fig. 3.5 Graph showing adjusted retention volume *vs.* carbon number for a set of series of homologous compounds containing various functional groups. R is a n-alkyl group.
(Reproduced, with permission, from *Analyt. Chem.*, 1972, **44**, p. 58)

Biochemical and Biological Materials

P.g.c. has been used on biochemical and biological materials, to give 'fingerprint' pyrograms, quantitative analyses, and information on the structure of biomacromolecules.

The literature on biological macromolecules to 1967 has been reviewed[6] (some work on p.g.c. of natural fibres will be found in the references cited under 'Textile Fibres' in the section on synthetic polymers). Stack pointed out that microgram-scale samples are required to minimise secondary reactions, but in some studies much larger samples have been used, limiting the possibility of reproducing the results and reducing their value in determining the initial breakdown mechanisms involved.

Amino-acids and the amino-acid moieties of proteins have received much attention, often to aid the elucidation of protein structure, and a very wide range of pyrolysis temperatures have been employed (less than 300 °C to greater than 800 °C). Winter and Albro[104] used the 'amine profile' obtained by pyrolysis of amino-acids to identify the parent acids, and found that the amines obtained on pyrolysis of proteins could also be related back to the constituent amino-acids. Simon *et al.*,[105,106] using only $3 \mu g$ samples, found that the amino-acid fragments caused by pyrolysis of peptides varied in proportion to the peptide linkages. In a further study,[107] each of seventeen amino-acids was found to produce at least one unique compound (the compounds had a variety of functional groups, and were identified by m.s.). It was concluded that the sequence in a peptide chain might be deduced from the pyrolysis product distribution, which varied qualitatively, as opposed to the purely quantitative variation observed by Simon *et al.* Stack[108a] differentiated the proteins from a variety of sources (but in particular dental enamel and dentin) by p.g.c., using differences of peak area in the pyrograms to discriminate between the samples. In this way, changes in the organic matrix of adult enamel were detected in enamel attacked by dental caries. Six amino-acids were characterised, using a ruby laser as pyrolyser;[68] under these conditions the unique products were found to be aldehydes. The variety of 'unique products' obtained from these studies of amino-acid pyrolyses shows the lack of uniformity in pyrolysis and/or g.c. conditions but does not imply lack of reproducibility of the various methods used. The yield of hydrogen cyanide, based on amino-nitrogen, varies with molecular structure; thus this single pyrolysis product, measured quantitatively, can give information about its parent amino-acid.[108b] The sequence distribution of amino-acids in actinomycins was found by the identification of the diketopiperazines formed from adjacent amino-acid pairs on pyrolysis.[108c]

An unusual p.g.c. system has been used for carbohydrates, polypeptides, and lipids in dilute aqueous solutions, employing a nickel-packed furnace and steam as the carrier gas[109] (as pointed out in Chapter 2, this unit is likely to induce catalytic reactions). Best discrimination was obtained with a pyrolysis temperature of 800 °C. When the organic material was present in trace concentrations, sample sizes as large as 0.25 ml were used.

Work on some cellulosic materials can be found in the references to textile fibres[18,19a] and in the review by Shafizadeh.[110] Martin and Ramstad,[111] using two G.C. columns in parallel, separated the compounds of lower molecular weight that were produced by radiative pyrolysis of cellulose.

Purines and pyrimidines were identified by p.g.c.; where the compounds had closely related structures, it was necessary to consider relative peak heights to differentiate the pyrograms.[112]

Ribonucleosides, ribonucleotides, and dinucleotides were characterised both quantitatively and in terms of microstructure by pyrolysis at 850 °C.[113] Ribonucleotide and ribonucleoside pyrograms correlate with those of the corresponding free bases and D-ribose. It was necessary to compare peak area ratios of the nitrogen-containing products to discriminate between parent bases (*c.f.* ref. 112).

A series of biochemical materials were identified both qualitatively and quantitatively by Stanford.[114] The substances investigated include atropine, haemin, bilirubin, vitamin A, n- and iso-leucine, vitamin B_{12}, and *p*-chloromercuribenzoic acid. It was found necessary to pyrolyse the samples in glass capillaries held in the coil of a filament rather than with the samples directly on a filament if quantitative results were to be obtained; this is the reverse of the normally recommended procedure with synthetic polymers. A series of twenty-one alkaloids were distinguished with a filament pyrolyser operated at 1000—1200 °C by comparing the pyrograms mainly in terms of the lower hydrocarbons produced.[115] The significance of the discrimination was shown by a statistical study of the method as applied to these alkaloids.

Several choline derivatives were identified and quantified on a nanogram scale by a p.g.c. method which produced and separated the dimethylamine derivatives.[116] For the examination of tissue extracts containing choline derivatives, preliminary purification steps are required (as is, of course, true for most p.g.c. work).

Vegetable and animal oils are frequently analysed by a g.c. method requiring a relatively long work-up. Janák[117] identified olive, hardened olive, and coconut oils by the C_1—C_6 hydrocarbons produced on pyrolysis. An unusual forensic application of p.g.c. was reported by Lloyd and Roberts[118] in a murder case. The stomach contents of the victim consisted entirely of recently consumed potato chips, and from this the cooking fat was extracted. Samples of fats from all fish-and-chip shops close to the point where the deceased was last seen alive were taken for comparison. The fats were hydrolysed to form tetramethylammonium salts of their constituent fatty acids in a single stage, so that small quantities could be used. These salts were pyrolysed in a Curie-point unit and the fats were typed according to their oleate-plus-stearate : palmitate peak-height ratios. Twenty-three possible origins were thus narrowed to three.

Antibiotics were qualitatively identified and structural characteristics investigated by p.g.c.[119] Antibiotics which are difficult to characterise by paper and thin-layer chromatography were used, and one was quantified in the presence of another by using selected pyrogram peaks. Both high- (600—1300 °C) and low-temperature (200—400 °C) pyrolyses were used in this study. A similar combination of high- and low-temperature pyrolyses was used to differentiate polyene antifungal antibiotics,[120] although low-temperature (380 °C) pyrolysis produced a more characteristic pyrogram.

Recent work on the characterisation of bacteria by p.g.c. has shown the potential and the drawbacks of the technique for such samples. Reiner[121] analysed over 1500 samples, mainly bacteria, and found that differences were frequently only to be found in peak areas; hence careful quantitation was required. Oyama and Carle[122a] showed that organisms of one species grown on different media gave different pyrograms; hence culture conditions must be carefully standardised. Again, in some pyrograms only the ratios of peak areas varied. Nevertheless, the method can distinguish normal and pathological cells,[122b] micro-organisms,[122c,d] and a large number of closely related species.[123] A completely automatic comparison of pyrograms of micro-organisms by a computer system was recently developed,[124] using results based on the p.g.c. system of Reiner.

. Where differences in pyrograms from species to species are small, highly reproducible pyrolysis systems will be required for interlaboratory comparison of results. A Curie-point pyrolyser has recently been designed to fill this requirement.[125]

References for Chapter 3

1 Levy, R. L., *Chromatog. Rev.*, 1966, **8**, 48.
2 Brauer, G. M., 'Techniques and Methods of Polymer Evaluation,' Vol. 2, Marcel Dekker, New York, 1970, p. 41.
3 Jones, C. E. R., and Reynolds, G. E. J., *Brit. Polymer J.*, 1969, **1**, 197.
4 Audebert, R., *Ann. Chim. (France)*, 1968, **3**, 49.
5 Jacqué, L., *Chimie et Industrie*, 1968, **100**, 1108.
6 Stack, M. V., in 'Gas Chromatography 1968,' ed. Harbourn, C. L. A., and Stock, R., Institute of Petroleum, London, 1969, p. 109.
7 Jones, C. E. R., and Reynolds, G. E. J., *Reports Progr. Appl. Chem.*, 1969, **54**, 518.
8 Perry, S. G., *Adv. Chromatog.*, 1969, **7**, 221.
9 Sarner, S. F., *J. Chromatog. Sci.*, 1972, **10**, 65.
10 Nelson, D. F., Yee, J. L., and Kirk, P. L., *Microchem. J.*, 1962, **6**, 225.
11 Cox, B. C., and Ellis, B., *Analyt. Chem.*, 1964, **36**, 90.
12 Groten, B., *Analyt. Chem.*, 1964, **36**, 1206.
13 Berton, A., *Chim. Analyt.*, 1965, **47**, 502.
14 Sadowski, F., and Kühn, E., *Farbe Lack*, 1963, **69**, 267.
15 Esposito, G. G., *Analyt. Chem.*, 1964, **36**, 2183.
16 Jain, N. C., Fontan, C. R., and Kirk, P. L., *J. Forensic Sci. Soc.*, 1965, **5**, 102.
17 (a) Wadsworth, M. J., UKAEA AWRE SSCD Memo 166 (1965); (b) Heinonen, K., and Komsi, I., *Kemian Teollisuus*, 1970, **27**, 631; (c) Stewart, W. D., *J. Forensic Sci.*, 1974, **19**, 121; (d) Wheals, B. B., and Noble, W., *Chromatographia*, 1972, **5**, 553.
18 Haase, H., and Rau, J., *Melliand Textilber.*, 1966, **47**, 434.
19 (a) Janiak, R. A., and Damerau, K. A., *J. Crim. Law, Criminology and Police Sci.*, 1968, **59**, 434; (b) Bortniak, J. P., Brown, S. E., and Sild, E. H., *J. Forensic Sci. Soc.*, 1971, **16**, 380.
20 Foxton, A. A., Hillman, D. E., and Mears, P. R., *J. Inst. Rubber Ind.*, 1969, 179.
21 (a) Fisher, G. E., and Neerman, J. C., *Ind. and Eng. Chem. (Product Res. and Development)*, 1966, **5**, 288; (b) Noble, W., Wheals, B. B., and Whitehouse, M. J., *Forensic Sci.*, 1974, **3**, 163; (c) Kimber, R. W. L., and Searle, P. L., *Geoderma*, 1970, **4**, 47; (d) Bracewell, J. M., Robertson, G. W., and Tate, K. R., *ibid.*, 1976, **15**, 209; (e) Bracewell, J. M., and Robertson, G. W., *J. Soil Sci.*, 1976, **27**, 196; (f) De Forest, P. R., and Kirk, P. L., *Criminologist*, 1973, **8**, 35; (g) Cope, J. F., *Analyt. Chem.*, 1973, **45**, 562
22 Quigley, D. A., Davies, D. G., and Evans, H. L., *Lab. Practice*, 1969, **18**, 421.
23 Nematollahi, J., Guess, W., and Autuan, J., *Microchem. J.*, 1970, **15**, 53.
24 Chiu, J., *Analyt. Chem.*, 1968, **40**, 1516.
25 Barlow, A., Lehrle, R. S., and Robb, J. C., *Polymer*, 1961, **2**, 27.
26 Voigt, J., *Kunststoffe*, 1965, **55**, 458.
27 Braun, D., and Vorendohre, G., *Farbe Lack*, 1963, **69**, 820.
28 Noffz, D., and Pfab, W., *Z. analyt. Chem.*, 1967, **228**, 188.
29 Noffz, D., Benz, W., and Pfab, W., *Z. analyt. Chem.*, 1968, **235**, 121.
30 McCormick, H., *J. Chromatog.*, 1969, **40**, 1.
31 Zizin, V. G., Berdina, L. Kh., and Avdeeva, M. P., *Zavodskaya Lab.*, 1970, **36**, 1307 (English translation).
32 Alekseeva, K. V., Khramova, L. P., and Strel'nikova, I. A., *Zavodskaya Lab.*, 1970, **36**, 1304 (English translation).
33 Tsuge, S., Okumoto, T., and Takeuchi, T., *J. Chromatog. Sci.*, 1969, **7**, 250.
34 Armitage, F., *J. Chromatog. Sci.*, 1971, **9**, 245.
35 Perry, S. G., *J. Gas Chromatog.*, 1967, **5**, 77.
36 (a) Shono, T., and Shinra, K., *Analyt. Chim. Acta*, 1971, **56**, 303; (b) Krishen, A., and Tucker, R. G., *Analyt. Chem.*, 1974, **46**, 29.
37 Dübler, K.-H., and Hagen, E., *Plaste Kautschuk*, 1969, **16**, 169.
38 Barlow, A., Lehrle, R. S., Robb, J. C., and Sunderland, D., *Polymer*, 1967, **8**, 537.
39 Bagby, G., Lehrle, R. S., and Robb, J. C., *Polymer*, 1969, **10**, 683.

40 Bell, F. A., Lehrle, R. S., and Robb, J. C., *Polymer*, 1971, **12**, 579.
41 (a) Guyot, A., Bert, M., Michel, A., and McNeil, I. C., *European Polymer J.*, 1971, **7**, 471; (b) Ferlanto, E. C., Lindemann, M. K., Lucchesi, C. A., and Gaskill, D. R., *J. Appl. Polymer Sci.*, 1971, **15**, 445.
42 Okumoto, T., Takeuchi, T., and Tsuge, S., *Bull. Chem. Soc. Japan*, 1970, **43**, 2080.
43 O'Mara, M. M., *J. Polymer Sci.*, *Part A-1*, 1970, **8**, 1887; *ibid.*, 1971, **9**, 1387.
44 Tanaka, M., Nishimura, F., and Shono, T., *Analyt. Chim. Acta*, 1975, **74**, 119.
45 Willmot, F. W., *J. Chromatog. Sci.*, 1969, **7**, 101.
46 Deur-Šiftar, D., and Švob, W., *J. Chromatog.*, 1970, **51**, 59.
47 Galvin-Vacherot, M., Eustache, H., and Pham Quang Tho, *European Polymer J.*, 1969, **5**, 211.
48 Galvin-Vacherot, M., *European Polymer J.*, 1971, **7**, 1455.
49 Thompson, R. N., Nau, C. A., and Lawrence, C. H., *Amer. Ind. Hygiene Assoc. J.*, 1966, 488.
50 Ney, E. A., and Heath, A. B., *J. Inst. Rubber Ind.*, 1968, **2**, 276.
51 Krishen, A., *Analyt. Chem.*, 1972, **44**, 494.
52 (a) Burns, D. T., Johnson, E. W., and Mills, R. F., *J. Chromatog.*, 1975, **105**, 43; (b) Takeuchi, T., Tsuge, S., and Okumoto, T., *J. Chromatog. Sci.*, 1968, **6**, 542.
53 Kirret, O., and Küllik, E., *Eesti NSV Teaduste Akad. Toimetised, Ködi Keem., Geol.*, 1969, **18**, 211.
54 Tsuge, S., Okumoto, T., Sagimura, Y., and Takeuchi, T., *J. Chromatog. Sci.*, 1969, **7**, 253.
55 Tsuge, S., Okumoto, T., and Takeuchi, T., *Macromol. Chem.*, 1969, **123**, 123.
56 Tsuge, S., Okumoto, T., and Takeuchi, T., *Bull. Chem. Soc. Japan*, 1969, **42**, 2870.
57 Tsuge, S., Okumoto, T., and Takeuchi, T., *Macromolecules*, 1969, **2**, 200; Tsuge, S., Ito, H., and Takeuchi, T., *Bull. Chem. Soc. Japan*, 1970, **43**, 3341.
58 Senoo, H., Tsuge, S., and Takeuchi, T., *J. Chromatog. Sci.*, 1971, **9**, 315.
59 Farré-Ruis, F., and Guiochon, G., *J. Gas Chromatog.*, 1967, **5**, 457.
60 Garzo, G., and Pehrsson, K., *J. Organometallic Chem.*, 1971, **30**, 187.
61 Knight, G. J., *J. Polymer Sci.*, *Part B*, 1967, **5**, 855.
62 Jones, C. E. R., and Reynolds, G. E. J., *J. Gas Chromatog.*, 1967, **2**, 25.
63 Hagen, E., *Plaste Kautschuk*, 1966, **13**, 140.
64 Zulaica, J., and Guiochon, G., *Analyt. Chem.*, 1963, **35**, 1724.
65 Liddicoet, T. H., and Smithson, L. H., *J. Amer. Oil Chemists' Soc.*, 1965, **42**, 1097.
66 Siggia, S., and Whitlock, C. R., *Analyt. Chem.*, 1970, **42**, 1719.
67 Nakagawa, T., Miyajima, K., and Uno, T., *J. Chromatog. Sci.*, 1970, **8**, 261.
68 Kojima, T., and Morishita, F., *J. Chromatog. Sci.*, 1970, **8**, 471.
69 Janàk, J., in 'Gas Chromatography 1960,' ed. Scott, R. P. W., Butterworths, London, 1960, p. 367.
70 Nelson, D. F., and Kirk, P. L., *Analyt. Chem.*, 1962, **34**, 899.
71 Nelson, D. F., and Kirk, P. L., *Analyt. Chem.*, 1964, **36**, 875.
72 Abbot, S. D., Hall, R. C., and Giam, C. S., *J. Chromatog.*, 1969, **45**, 317.
73 Schmidt, D. E., Szilagyi, P. I. A., and Green, J. P., *J. Chromatog. Sci.*, 1969, **7**, 248.
74 Legate, C. E., and Burnham, H. D., *Analyt. Chem.*, 1960, **32**, 1042.
75 Perry, S. G., *J. Gas Chromatog.*, 1964, **2**, 93.
76 Romovàcek, J., and Kubát, J., *Analyt. Chem.*, 1968, **40**, 1119.
77 Leplat, P., *J. Gas Chromatog.*, 1967, **5**, 128.
78 Poxon, D. W., and Wright, R. G., *J. Chromatog.*, 1971, **61**, 142.
79 Biscar, J. P., *J. Chromatog.*, 1971, **56**, 348.
80 Levy, R. L., Wolf, C. J., Grayson M. A.. Gibert, J., Gelpi, E., Updegrove, W. S., Zlatkis, A., and Oro, J., *Nature*, 1970, **227**, 148.
81 Levy, R. L., and Wolf, C. J., *J. Chromatog. Sci.*, 1970, **8**, 524.
82 Oró, J., Updegrove, W., McReynolds, J., Ibanez, J., Gil-Av, E., Flory, D., and Zlatkis, A., *J. Chromatog. Sci.*, 1970, **8**, 297.

83 Oró, J., Updegrove, W. S., Gibert, J., McReynolds, J., Ibanez, J., Gil-Av, E., Zlatkis, A., Flory, D. A., Levy, R. L. and Wolf, C., *Science*, 1970, **167**, 765.
84 Simmonds, P. G., Schulman, G. P., and Stembridge, C. H., *J. Chromatog. Sci.*, 1969, **7**, 36.
85 Sternberg, R. C., Krull, I. H., and Friedel, G. D., *Analyt. Chem.*, 1966, **38**, 1639.
86 Keulemans, A. I. M., and Perry, S. G. in, 'Gas Chromatography 1962,' ed. van Swaay, M., Butterworths, London, 1962, p. 356.
87 Holman, R. T., Deubig, M., and Hayes, H., *Lipids*, 1966, **1**, 247.
88 Cramers, C. A. M. G., and Keulemans, A. I. M., *J. Gas Chromatog.*, 1967, **5**, 58.
89 Brown, R. A., *Analyt. Chem.*, 1971, **43**, 900.
90 Kossiakoff, A., and Rice, F. O., *J. Amer. Chem. Soc.*, 1943, **65**, 590.
91 Fanter, D. L., Walker, J. Q., and Wolf, C. J. *Analyt. Chem.*, 1968, **40**, 2168.
92 Gough, T. A., and Walker, E. A., *J. Chromatog. Sci.*, 1970, **8**, 134.
93 (a) Van Cauwenberghe, K., Vandewalle, M., and Verzele, M., *J. Chromatog. Sci.*, 1969, **7**, 698; (b) Svob, V., Deur-Šiftar, D., and Cramers, C. A., *Chromatographia*, 1972, **5**, 540.
94 Groenendyk, H., Levy, E. J., and Sarner, S. F., *J. Chromatog. Sci.*, 1970, **8**, 115.
95 Walker, J. Q., and Maynard, J. B., *Analyt. Chem.*, 1971, **43**, 1548.
96 Dhont, J. H., *Nature*, 1963, **200**, 882.
97 Levy, E. J., and Paul, D. G., *J. Gas Chromatog.*, 1967, **5**, 136.
98 Walker, J. Q. and Wolf, C. J., *Analyt. Chem.*, 1968, **40**, 711.
99 Wolf, T., and Rosie, D. M., *Analyt. Chem.*, 1967, **39**, 725.
100 (a) Groenendyk, H., Levy, E. J., and Sarner, S. F., *J. Chromatog. Sci.*, 1970, **8**, 599; (b) Merritt, C., and Di Pietro, C., *Analyt. Chem.*, 1972, **44**, 57.
101 Liebman S. A., Ahlstron, D. H., Creighton, T. C., Pruder, G. D., Averitt, R., and Levy, E. J., *Analyt. Chem.*, 1972, **44**, 1411.
102 Denker, W. D., and Wolf, C. J., *J. Chromatog. Sci.*, 1970, **8**, 534.
103 Dhont, J. H., *Analyst*, 1964, **89**, 73.
104 Winter, L. N., and Albro, P. W., *J. Gas Chromatog.*, 1964, **2**, 1.
105 Simon, W., and Giacobbo, H., *Angew. Chem. Internat. Edn.*, 1965, **4**, 938.
106 Voellmin, J. A., Kriemler, P., Omura, I., Seibl, J., and Simon, W., *Microchem. J.*, 1966, **11**, 73.
107 Merritt, C., and Robertson, D. H., *J. Gas Chromatog.*, 1967, **5**, 96.
108 (a) Stack, M. V., *J. Gas Chromatog.*, 1967, **5**, 22; (b) Johnson, W. R., and Kang, J. C., *J. Org. Chem.*, 1971, **36**, 189; (c) Mauger, A. B., *Chem. Comm.*, 1971, 39.
109 Lysgi, I., and Nelson, K. H., *Analyt. Chem.*, 1968, **40**, 1365.
110 Shafizadeh, F., *Adv. Carbohydrate Chem.*, 1968, p. 419.
111 Martin, S. B., and Ramstad, R. W., *Analyt. Chem.*, 1961, **33**, 982.
112 Jennings, E. C., and Dimick, K. P., *Analyt. Chem.*, 1962, **34**, 1543.
113 Turner, L. P. and Barr W. R., *J. Chromatog. Sci.*, 1971, **9**, 176.
114 Stanford, F. G., *Analyst*, 1965, **90**, 266.
115 Kingston, C. R., and Kirk, P. L., *Bull. Narcotics*, 1965, **17**, 19.
116 Szilagyi, I. A., Schmidt, D. E., and Green, J. P., *Analyt. Chem.*, 1968, **40**, 2009.
117 Janák, J., *Nature*, 1960, **185**, 684.
118 Lloyd, J. B. F., and Roberts, B. R. G., Home Office C.R.E. (Aldermaston) Circulation, November 1970.
119 Brodasky, T. F., *J. Gas Chromatog.*, 1967, **5**, 311.
120 Burrows, H. J., and Callam, D. H., *J. Chromatog.*, 1970, **53**, 566.
121 Reiner, E., *J. Gas Chromatog.*, 1967, **5**, 65.
122 (a) Oyama, V. E., and Carle, G. C., *J. Gas Chromatog.*, 1967, **5**, 151; (b) Reiner, E., and Hicks, J. J., *Chromatographia*, 1972, **5**, 525; (c) Derenback, J. B., and Ehrhardt, M., *J. Chromatog.*, 1975, **105**, 339; (d) Quinn, P. A., *J. Chromatog. Sci.*, 1974, **12**, 796.
123 Reiner, E., Hicks, J. J., Ball, M. M., and Martin, W. J., *Analyt. Chem.*, 1972, **44**, 1058.
124 Menger, F. M., Epstein, G. A., Goldberg, D. A., and Reiner, E., *Analyt. Chem.*, 1972, **44**, 423.
125 Meuzelaar, H. L. C., and in't Veld, R. A., *J. Chromatog. Sci.*, 1972, **10**, 213.

4 Identification of Peaks

The main initial use of pyrolysis–gas chromatography was for the identification of an unknown substance by comparison with a standard or a series of standards. The pyrograms of most organic materials are of comparable complexity to an i.r. spectrum, and it is possible to use the pyrograms purely as a pattern or 'fingerprint' to say that two materials are similar or different.[1] When faced with a problem of identifying an unknown material, the answer can in many cases be rapidly supplied by p.g.c. Groten, in 1964, built up a library of the pyrograms of over 150 polymers. He used a filament pyrolyser and analysed the pyrolysates on a 12 foot column of 20% carbowax on Diatoport P.[2] He showed that all the polymers could be distinguished by their pyrograms, purely on the basis of pattern recognition. The same approach to the identification of paint binders was used by Sadowski and Kuhn,[3] and by Jain, Fontan, and Kirk,[4] both groups of workers using didecyl phthalate as the stationary phase. Collections of fingerprint pyrograms for textile fibres have been published by Gocken and Cates,[5] who used a furnace pyrolyser, an Apiezon L column, and who programmed the temperature from 50 to 225 °C at a rate of 10 °C min[-1]. This work was followed by that of Janiak and Damerau,[6] who used a filament pyrolyser and a Carbowax 20M column programmed from 65 to 225 °C, and by that of Bortniak, Brown, and Sild,[7] who distinguished 15 groups of acrylic and modacrylic fibres on a Carbowax 1540 column programmed from 50 to 180 °C. Nelson and Kirk[8] showed that 27 barbiturates could be distinguished from each other. They used a resistance-heated platinum cup pyrolyser and separated the pyrolysates on a column of silicone oil. They published their pyrograms in the form of bar graphs, with the peaks replaced by a single line of corresponding height, as do Liddicoet and Smithson[9] in their work on the p.g.c. of surfactants. Since an identification is based on pattern recognition, and bar graphs remove a lot of the pattern, the value of this procedure is to be doubted.

To make valid comparisons of materials on the basis of fingerprint pyrograms, it is essential that the conditions be exactly reproduced for pyrolysis of sample and of control, as we show elsewhere. However, since each peak in the pyrogram represents a compound formed in the breakdown of the sample, there is a great deal of information to be gained from knowing what these compounds are. With this information it is not necessary to reproduce another worker's chromatographic conditions to be able to use his pyrogram, as it is possible to relate one to the other. It is then possible to concentrate on peaks of genuine structural significance rather than on the series of hydrocarbons which are common to the pyrolysates of very many materials, and which can give a superficial similarity to pyrograms of different materials. Stationary phases most suitable for the compounds involved can be used.

A number of workers have identified the peaks in their pyrograms. The methods used are based on measurement of retention times, trapping of the eluates followed by spectroscopic methods, chemical modification (*e.g.* hydrogenation of the pyrolysate), and linked G.C.–M.S. systems.

Identification from Retention Times

The methods of identification based on retention times usually involve comparing the unknowns with standard samples of compounds whose presence is thought to be likely from a consideration of the structure of the sample, or from the results of large-scale pyrolysis experiments.

Using the method of comparing the retention times of a series of likely standards with those of their pyrolysate peaks, Lehman and Brauer[10] identified the peaks in the relatively uncomplicated pyrograms of polystyrene and poly-(methyl methacrylate). They used 3 columns, and thus confirmed their identifications. These contained Apiezon L, Chlorowax 70, and dinonyl phthalate. They also used a column of silica gel to separate the light hydrocarbons and CO_2. Also using this method, Braun and Vorendohre[11] identified 39 peaks in the pyrograms of a large number of polymers and copolymers. These included polystyrene, poly-(methyl methacrylate), polyacrylonitrile, poly(vinyl acetate), poly(vinyl alcohol), poly(vinyl chloride), polyethylene, and polypropylene. The pyrolysates were separated on a 2 m column of di(ethylhexyl) sebacate, and the retention times of the identified peaks are listed. Unfortunately, they do not quote the percentage of stationary phase in the column.

To get some idea of what standards to try out, it is often useful to pyrolyse a large quantity of the material and isolate the different breakdown products. The pyrolysis under these conditions is not likely to proceed in the same manner as it would when micro quantities are being used, but some correlation should be possible. This of course assumes that there is plenty of sample. This course was adopted by Nelson and Kirk[12] to identify the pyrolysis products of barbiturates. They heated 3 g of barbiturate in a test-tube and collected the products *via* a condenser. The products were then separated on a preparative-scale chromatograph, identified by standard methods (including i.r.), and then used as standards of known retention times for identifying the peaks in the pyrograms.

When Winter and Albro[13] studied the pyrolysis of amino-acids and proteins, they knew from previous work that the main pyrolysis products were likely to be ammonia and aliphatic amines. They therefore determined the retention volumes of the likely amines relative to that of tripropylamine for comparison with the peaks in the pyrograms. They used a katharometer detector and connected the outlet to a cell containing Methyl Red or Nessler's Reagent to confirm that the eluate was an amine, this nicely confirming the evidence of the retention volume.

Cramers and Keulemans[14] used retention times to identify the fragments in their studies on the pyrolysis of hydrocarbons. They used a 30 m capillary column coated with dimethyl sulpholane and compared the retention times of the unknowns with the retention data of some 300 hydrocarbons which they had obtained in their laboratory.

When one is using retention times for identification of peaks, it is helpful to have some prior knowledge of what compounds are likely to occur. For this reason it is not a very useful approach when one has a small quantity of a completely unknown sample. To be at all certain of the identification it is essential to check the retention times on at least two independent stationary phases.

Identification by Chemical Means

It is possible to simplify the pyrogram as an aid to the subsequent identification of the peaks from their retention times by chemical modification of the pyrolysate. Several workers have simplified the pattern of hydrocarbon peaks obtained from the pyrolysis of polyethylene by hydrogenation. The pyrogram consists of a series of triplets (Figure 4.1). These were identified[15,16] by using hydrogen as carrier gas and passing the stream through a precolumn containing a hydrogenation catalyst. This reduced the pyrogram to a series of n-paraffins, identifiable as such by their retention times, which also correspond to one peak in each triplet. Hence the other two peaks in the triplet were n-olefins, since they were reduced to n-paraffins on hydrogenation. Since they were formed by pyrolysing a saturated hydrocarbon polymer, they were likely to be α-olefins and α,ω-olefins; such species were identified by trapping the C_{12} triplet in a solution of potassium permanganate in glacial acetic acid, when the double bonds were oxidised to carboxy-groups. The n-dodeca-1-ene gave undecanoic acid and the n-dodeca-1,11-diene gave sebacic acid. These acids were methylated with diazomethane, and the esters then identified gas chromatographically from their retention times. Gough and Walker[17] also used hydrogenation to simplify the pyrograms from normal and branched aliphatic hydrocarbons. They found that, by so doing, isomerisation of olefins was avoided, thus making the pyrograms more reproducible.

Identification by Trapping followed by Spectroscopy

A direct method of identifying peaks is to trap them as they emerge from the column and then identify them by chemical means, as did Kolb *et al.* with α,ω-olefins (*vide supra*), or by spectroscopic means. For this, one uses a katharometer detector, which is non-destructive, or else one places a splitter in the effluent stream, sending a small portion to the flame ionisation detector and the greater portion to the trap. Because the quantity of material in a single pyrogram peak is so small, up to a dozen samples may have to be run, the required peak being trapped each time. A simple trapping technique is that described by Curry *et al.*,[18] in which the peak is trapped in a small glass phial cooled in liquid nitrogen. These authors describe the preparation of a micro-KBr disc from which an i.r. spectrum may be obtained from as little as 1 microgram of material. The peaks may also be conveniently trapped in a short capillary tube containing a small quantity of column packing or other adsorbent.[19-21] Fisher and Neerman[22] examined brake-lining materials and identified the aromatic hydrocarbons and phenols in their pyrograms by trapping them in a capillary tube and then identifying them from their i.r. spectra.

Daniel and Michel,[23] working on the pyrolysis of vinyl acetate and its copolymers with other esters, identified the peaks by two i.r. methods. The first in-

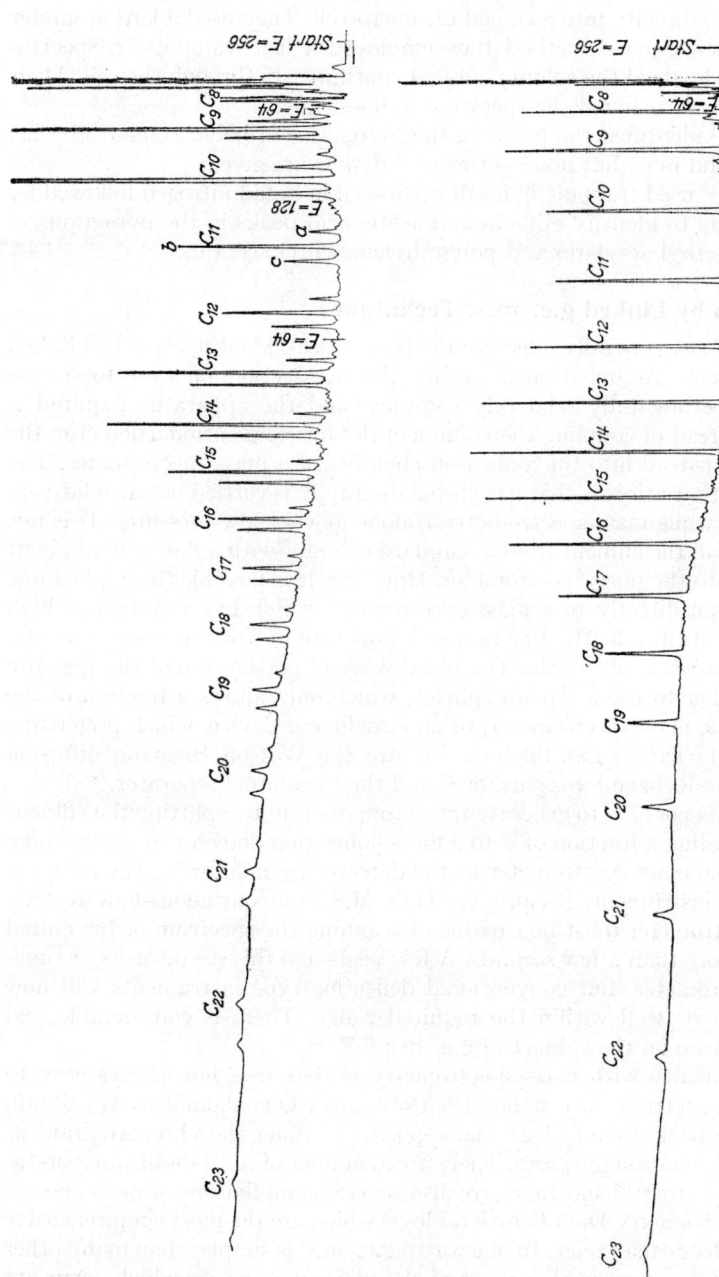

Fig. 4.1 Pyrograms of low-pressure polyethylene without hydrogenation (upper) and after hydrogenation (lower pyrogram). In the upper trace each triplet consists of peaks for (a) the paraffin, (b) the α-olefin, and (c) the α,ω-diolefin corresponding to the carbon number shown.
(Reproduced, with permission, from Z. analyt. Chem., 1967, **228**, p. 192).

volved trapping directly into a cooled i.r. micro-cell. They used a katharometer detector. In their second method they employed a fast-scanning i.r. spectrophotometer and passed the column effluent continuously through the cell. Their spectrophotometer scanned the spectrum in 6 seconds.

Tsuge et al.[24] identified the peaks in the pyrograms of polycarbonates by i.r. spectroscopy and m.s., but no experimental details are given.

Barrall et al.[25] used trapping in a salt micro-cell in liquid nitrogen followed by i.r. spectroscopy to identify ethanol and acetic acid peaks in the pyrograms of poly(ethylene–ethyl acrylate) and poly(ethylene–vinyl acetate).

Identification by Linked g.c.–m.s. Techniques

Probably the most generally effective method of peak identification is a linked g.c.–m.s. system. As mentioned earlier (in the section on detectors), the method is experimentally relatively complex, and the apparatus required is expensive. Instead of coupling the column outlet to a conventional detector, the effluent is led instead into the ionisation chamber of a mass spectrometer. The first difficulty that arises is that gas chromatography is carried out at relatively high pressure, while mass spectrometry is done at very low pressures. It is not possible to let all the effluent from a standard column (with a flow rate of about $60\,ml\,min^{-1}$) into the mass spectrometer. However, it is possible to couple a fine capillary column directly to a mass spectrometer which has a system of high pumping capacity (i.e. $300\,l\,s^{-1}$ or more). A flow rate of 2 or $3\,ml\,min^{-1}$ into the ionisation chamber is allowable. The usual ways of getting round the pressure barrier are either to use a stream splitter, which only allows a fraction of the effluent into the mass spectrometer, or an enrichment device, which preferentially removes the carrier gas. Such devices are the Watson–Biemann diffusion separator,[26] the Ryhage jet separator,[27] and the membrane separator.[28]

Although it is possible to generate the chromatogram by splitting the effluent stream and leading a fraction of it to a flame ionisation detector, it is now more usual to use the mass spectrometer as the detector by monitoring the total ion current in the instrument. Because the G.C.–M.S. is a continuous-flow system, the mass spectrometer must be capable of scanning the spectrum of the eluted peaks in no more than a few seconds. A few years ago this meant using a time-of-flight spectrometer, but conventional deflection-type instruments will now scan over periods well within the required range. There is considerable and growing literature on the subject of g.c.–m.s.[27,29–35]

Another difficulty with mass spectrometry is that it is not always easy to interpret the spectrum once it has been obtained. One should always obtain spectra of the background at as many positions along the chromatogram as possible, for elimination purposes. There are a number of texts on the interpretation of mass spectra[36,37] and there are also several compilations of mass spectra (e.g. Mass Spectrometry Data Centre tables,[38] which are the most comprehensive so far) to which one may refer. In one way p.g.c.–m.s. is simpler than many other applications of g.c.–m.s. in that most of the molecular species which occur are fairly simple, and the spectra may be readily recognised by inspection or by reference to the tables. Homologous series often occur in pyrograms, making the higher members more readily identifiable from a knowledge of the spectra of the lower members already identified.

References for Chapter 4

1 Kirk, P. L., *J. Gas Chromatog.*, 1967, **1**, 11.
2 Groten, B., *Analyt. Chem.*, 1964, **36**, 1206.
3 Sadwoski, F., and Kuhn, E., *Farbe Lack*, 1963, **69**, 267.
4 Jain, N. C., Fontan, C. R., and Kirk, P. L., *J. Forensic Sci. Soc.*, 1965, **5**, 102.
5 Gocken, U., and Gates, D. M., *Appl. Polymer Symposium*, 1966, **2**, 15.
6 Janiak, R. A., and Damerau, K. A., *J. Crim. Law, Criminology and Police Science*, 1965, **59**, 434.
7 Bortniak, J. P., Brown, S. E., and Sild, E. H., *J. Forensic Sci.*, 1971, **16**, 380.
8 Nelson, D. F., and Kirk, P. L., *Analyt. Chem.*, 1962, **34**, 899.
9 Liddicoet, T. N., and Smithson, L. H., *J. Amer. Oil Chemists' Soc.*, 1965, **42**, 1097.
10 Lehman, F. D., and Brauer, G. M., *Analyt. Chem.*, 1961, **33**, 673.
11 Braun, D., and Vorendohre, G., *Farbe Lack.*, 1963, **69**, 820.
12 Nelson, D. F., and Kirk, P. L., *Analyt. Chem.*, 1964, **36**, 875.
13 Winter, L. N., and Albro, P. W., *J. Gas. Chromatog.*, 1964, **1**, 1.
14 Cramers, C. A. M. G., and Keulemans, A. I. M., *J. Gas Chromatog.*, 1967, **5**, 58.
15 Kolb, B., and Kaiser, K. H., *J. Gas Chromatog.*, 1964, **2**, 233.
16 Kolb, B., Kemmner, G., Kaiser, K. H., Cieplinski, E. W., and Ettre, L. S., *Z. analyt. Chem.*, 1965, **209**, 302.
17 Gough, T. A., and Walker, E. A., *J. Chromatog. Sci.*, 1970, **8**, 134.
18 Curry, A. S., Read, J. F., Brown, C., and Jenkins, R. W., *J. Chromatog.*, 1968, **38**, 200.
19 Amy, J. W., Chait, E. M., Baitinger, W. E., and McLafferty, F. W., *Analyt. Chem.*, 1965, **37**, 1265.
20 Damico, J. N., Wong, N. P., and Sphon, J. A., *Analyt. Chem.*, 1967, **39**, 1045.
21 Cartwright, M., and Heywood, A., *Analyst*, 1966, **9**, 337.
22 Fisher, G. E., and Neerman, J. C., *Ind. and Eng. Chem. (Product. Res. and Development)*, 1966, **5**, 288.
23 Daniel, J. C., and Michel, J. M., *J. Gas Chromatog.*, 1967, **5**, 437.
24 Tsuge, S., Okumoto, T., Sugimura, Y., and Takeuchi, T., *J. Chromatog. Sci.*, 1969, **7**, 253.
25 Barrall, E. M., Porter, R. E., and Johnson, J. F., *Analyt. Chem.*, 1963, **35**, 73
26 Watson, J. T., and Biemann, K., *Analyt. Chem.*, 1965, **37**, 844.
27 Ryhage, R., *Analyt. Chem.*, 1964, **36**, 759.
28 Llewellyn, P. M., and Littlejohn, D. P., U.S.P. 3 471 692 (1969).
29 Ryhage, R., Wilkstrom, S., and Waller, G., *Analyt. Chem.*, 1965, **37**, 435.
30 McFadden, W. H., *Adv. Chromatog.*, 1967, **4**, 265.
31 Banner, A. E., Elliot, R. M., and Kelly, W., in 'Gas Chromatography 1964,' ed. Goldup, A., Institute of Petroleum, London, 1965, p. 180.
32 Hammar, C. G., *Acta Pharm. Suecica*, 1971, **8**, 129.
33 Hammar, C. G., and Hessling, R., *Analyt. Chem.*, 1971, **43**, 298.
34 Van Cauwenberghe, K., Vandewalle, M., and Verzele, M., *J. Gas Chromatog.*, 1968, **6**, 72.
35 Blumer, M., *Analyt. Chem.*, 1968, **40**, 1590.
36 Budzikiewicz, H., Djerassi, C., and Williams, D. H., 'Interpretation of Mass Spectra of Organic Compounds,' Holden Day, San Francisco, 1964.
37 McLafferty, F. W., 'Interpretation of Mass Spectra: An Introduction,' Benjamin, New York, 1966.
38 'Eight Peak Index of Mass Spectra,' Mass Spectrometry Data Centre, Aldermaston, Reading, England, 1970.

5 Standardisation in Pyrolysis–Gas Chromatography

In Chapter 2 we discussed the variety of pyrolyser types and those of their individual characteristics which can affect the nature of the pyrolysate produced. In the examples of applications of p.g.c. mentioned in Chapter 3, the range of pyrolysis temperatures and sample sizes that have been used were shown, and these will of course affect the nature of the pyrolysate. In this chapter we will discuss standardisation (or the lack of it) in the chromatographic system. The system will be considered as consisting of (*a*) the instrument, presumed to be of limited operational flexibility, and (*b*) the column packing and the operating conditions determined by the user.

Instruments

There are several companies which offer gas-chromatographic instrumentation. Most supply a range of instruments to cover the majority of g.c. applications. It is inevitable that several types of detector are offered, so that requirements of, for example, specific sensitivity to certain compounds (*e.g.* the electron-capture detector for halogenated, organometallic, conjugated carbonyl, nitrile, nitrate, and sulphur compounds) or linear response to particular concentration ranges, *etc.*, can be catered for. The best detector on which to standardise in p.g.c. is either that giving the highest response to the widest range of pyrolysates or, if no detector stands out in this respect, to the most widely used detector; fortunately, both criteria are to be found in the F.I.D.

Working back from the detector, a further variable is found in the wide range of column sizes commercially available, which could surely be reduced to a much smaller but still adequate range.

It is our opinion that, where many different types of polymer are to be subjected to p.g.c., it is necessary to use a temperature- or flow-programming device on the column to separate pyrolysis products where volatilities cover a wide range; unfortunately, programmers from different manufacturers rarely have identical ranges of programming rates.

It is not only the G.C. instrument itself which can introduce variables into p.g.c., but the use of different recorder chart speeds can also complicate the comparison of pyrograms, as is found in comparing i.r. spectra in the form of charts showing linear wavenumber and linear wavelength, respectively.

Column Packing and Operating Conditions

Probably the most confusing aspect of gas chromatography is the complex variety of packing materials available to the chromatographer. A range of solid supports, each obtainable with any one of several surface modifications, can be coated with any of literally hundreds of liquid phases in whatever proportion the user wishes. This makes it unlikely that independent workers will chance to use identical packings, making the comparison of fingerprint pyrograms very difficult; this is a problem encountered in many other aspects of chromatography.

The column temperature, and the composition and flow rate of the carrier gas, must be precisely defined if there is to be hope of reproducing a published set of pyrograms. If these operating conditions, the column packing, and its dimensions cannot be precisely reproduced, some useful comparative data may still be obtained if retention data are reproduced relative to some known eluate or eluates. This will be true in fingerprint pyrograms only if a few well-resolved peaks are present. When the pyrogram contains a large number of peaks with several overlapping, there is little hope of a successful comparison with a second pyrogram unless the g.c. conditions were identical for the production of both pyrograms, since the degree of overlap will vary with g.c. conditions. Nevertheless, for very simple pyrograms, relative retention data can aid identification.

A simple system for recording g.c. data was introduced by Evans and Smith.[1] In this system 'the theoretical nonane value' (R_{X9}), which is the ratio of the retention volume of the eluate under consideration to the retention volume of n-nonane, is found. This ratio is highly dependent on temperature, and its accuracy falls off as the difference in the two retention volumes increases.[2]

A system of more general application was introduced by Kovats,[3,4] who defined a retention index I_X for any eluate as follows:

$$I_X = \frac{100 \, [\log V_g(X) - \log V_g(P_z)]}{\log V_g(P_{z+1}) - \log V_g(P_z)} + 100 \, z$$

where $V_g(P_z) \leqq V_g(X) \leqq V_g(P_{z+1})$

$V_g(X)$, $V_g(P_z)$, and $V_g(P_{z+1})$ are the specific retention volumes of the compound X, the normal alkane P_z, and the next higher homologue P_{z+1}, respectively, where z represents the number of carbon atoms in P_z. As shown, P_z and P_{z+1} are chosen as those hydrocarbons eluted immediately prior to and after the compound X. The retention indices of the n-alkanes are arbitrarily defined as $100z$ for all columns under all stated conditions. The experimentally determined indices of all other compounds refer strictly to the g.c. conditions under which they were obtained, and will vary with the stationary phase, the more so the greater the difference in polarity of the substance and the n-alkane series. Kovat's index enables one to locate a compound with respect to the n-alkane series; for example, a compound with an index of 1240.6 is one which is eluted between n-dodecane and n-tridecane, 40.6 giving the precise location between the two hydrocarbons. The figure quoted is for a compound eluted from Apiezon L at 210 °C and the index is reported as $I_{210}^{\text{Apiezon}} = 1240.6$. Experimental error in such an index is said to be of the order of two index units.[4] It is necessary to quote the temperature since the index can vary with temperature, and this variation can be used as an aid in the identification of an eluate.[5,6]

In a system subjected to a temperature programme the n-alkanes are eluted roughly linearly with time, so that a new index can be defined, substituting elution temperatures for the $\log V_g$'s of the above equation.

Certain functional groups have been found to cause a fixed difference in index between compounds containing a particular functional group and the parent hydrocarbon series; this difference has been called the functional retention index.[7] Other classifications based on the difference in index of compounds from column to column when eluted on two different columns have also been made. Some seventy references have been quoted on this interesting subject of the relationships between molecular structure and retention indices.[8] Retention indices, then, provide a reliable means of reporting chromatographic data, and as with specific retention volumes, from which they are derived, many interesting correlations with molecular structure have been found, providing an aid to the identification of unknown eluates.

When the indices of known compounds are compared on different column packings, something of the characteristics of the packings can be discovered. Rohrschneider[9] used the retention indices of benzene, ethanol, methyl ethyl ketone, nitromethane, and pyridine as measures of the 'polarity' of stationary phases, the five compounds being chosen to cover a range of material of widely differing 'polarity' (*i.e.* covering a spectrum of electron-donor/-acceptor character) The retention indices of these compounds are determined on a squalane column and then on the column to be investigated; the differences in retention index for each compound (ΔI) are noted. All results must be obtained at a fixed temperature. These figures, or $I/100$, the Rohrschneider constants, can then be used in comparisons of different stationary phases. These constants[10] and ΔI values for a larger range of compounds,[11] mainly of higher boiling points than those chosen by Rohrschneider, have been determined for over 299 liquid phases. They demonstrate the similarity of many of the commercial liquid phases available, thus providing evidence which (hopefully) will lead to a reduction in the bewildering number of stationary phases in use. A small number of liquid stationary phases could then be recommended as suitable for the majority of g.c. separations, though no doubt certain specialised phases will always be required.

The seemingly impossible task of finding a stationary phase capable of performing all required separations suggests that in p.g.c. more than one column packing will have to be used if the full potential of the technique is to be realised. From the viewpoint of the analyst interested in obtaining 'fingerprint' pyrograms, a column giving well-shaped peaks for a wide range of possible pyrolysates is required, and this information cannot be obtained from the Rohrschneider constants. Perhaps the standardisation of g.c. techniques for p.g.c. is best considered at two levels: (*a*) 'fingerprinting', in which a small number (preferably one or two) of column packings are used for general work, as in the identification of a range of plastics and paint vehicles presented later in this chapter, and (*b*) specialised areas where unusual series of products are produced; for example, alkyl quaternary ammonium halides, which give pyrolysates consisting mainly of alkyl halides and amines.[12] It is hoped that each specialised area could be catered for by one stationary phase, the choice being aided by the availability of Rohrschneider constants and other data.

A Standard System

The previous sections of this chapter show the difficulties in standardising p.g.c., but a standard system for fingerprinting materials, backed by a collection of pyrograms of known substances, would greatly increase the value of p.g.c. for many laboratories. Unfortunately, no such system has gained acceptance. It may be that currently available equipment and materials cannot satisfactorily solve the problem for all pyrolysable materials. Our approach to the restricted problem of standardising the fingerprinting technique for polymers of mainly synthetic origins will now be described. Another system designed for a similar application to ours is under development.[13] In this other system the column and its operating conditions are considerably different from those of the system to be described later in this chapter; consequently, significant differences in the types of volatiles separated and eluted by the two g.c. systems are apparent. It is possible that the other system and that described in this chapter will be usefully complementary.

The aims of the work described here were to find a system of p.g.c. which would require a sample size of less than $20\,\mu g$ and could be used in the identification of a range of polymers, with particular emphasis on those commonly found in paints, and which also would be sufficiently reproducible that pyrograms could be duplicated in other laboratories. The p.g.c. method described here is sufficiently reproducible to allow identification of a sample by reference to a collection of pyrograms. A collection of pyrograms covering a range of polymers and paints is presented, together with an index based on peak heights and retention times. Because overlapping peaks are unavoidable in the pyrograms of some polymers, chromatographic conditions must be kept constant, and the identification of a pyrogram requires visual comparisons with the collection of pyrograms. The overlap of peaks renders an identification system based on retention indices (such as that described by Kazyak[14]) untenable.

Some polymers, such as the methacrylates, degrade almost entirely to their monomers. Others, such as the alkyds, break down on pyrolysis to yield a large number of fragments, many being very volatile. The ideal G.C. column for this work would therefore resolve volatile fractions such as methane and ethane and also elute much less volatile materials, styrene and phthalic anhydride being common examples. In addition, it must be capable of giving the same chromatographic performance when made up in different laboratories. In order to encompass the second requirement successfully, it was necessary to reduce the 'ideal' performance below that of a column satisfying the first requirement.

Seven liquid-phase packings and three solid-phase packings were investigated for this work, and only Porapak Q and Chromosorb 102 were found to be satisfactory in terms of reproducibility from column to column, useful lifetime, and ability to discriminate between polymer pyrolysates. The major portion of this work was therefore restricted to pyrolyses using one of these (Porapak Q) as the column packing. Fixed G.C. operating conditions were also used, and pyrograms capable of effective comparison with this collection must be made on a G.C.

containing columns of the same dimensions, packed with Porapak Q, and capable of the same temperature programme as used in this work.

The literature on p.g.c. is expanding, and there have been several recent reviews.[15-17] There are three main types of pyrolyser for solid samples: hot-filament, Curie-point, and furnace (described in detail in Chapter 2, together with other forms of heating, including electric discharge and laser light). Pyrolysis conditions must be kept constant if pyrograms are to be produced that can be usefully compared with the ones in this collection. After testing a hot-filament and a Curie-point pyrolyser, the latter was chosen, and three units of this type were used as part of the study of reproducibility. This Curie-point unit gives rapid and reproducible heating to the Curie point of the pyrolysis wire. If pyrolysis wires of a fixed composition are used in different pyrolysers, identical pyrolysis temperatures are obtained, giving highly reproducible pyrograms. A pyrolysis temperature of 610 °C was chosen, this being in the region where most polymers give characteristic fragmentation patterns.

In much of the early work on p.g.c., little attempt was made to identify the compounds forming the peaks in the pyrogram. The starting material was identified from its pyrogram solely on the basis of pattern recognition.[18] It is obvious that the utility of the pyrograms as a means of identification may be increased from a knowledge of what compounds are formed during pyrolysis. Attention may then be focussed on the key fragments of real structural signific-ance rather than, for instance, on the series of hydrocarbons which are common to many pyrolysates. Because most of the pyrograms in the literature were obtained using heated-filament- or furnace-type pyrolysers, often using samples as large as several milligrams, and because of the essential unreliability of identi-fications based on retention times, it was decided to identify the peaks in all the pyrograms forming our collection, using a linked P.G.C.–M.S. system as the method of choice for obtaining rapid, unequivocal identifications.

Experimental

The following column packings were investigated: Carbowax 20M, OV17, and Apiezon L, each 10% on acid-washed, silanised Chromosorb G; cyclohexane dimethanol succinate and SE30, both 10% on acid-washed, silanised Chromo-sorb W; E301, 10% on acid-washed Celite C; Durapak (Carbowax 400) and the solid phases Porasil C, Porapak Q, and Chromosorb 102. Each of these packings was tested to ascertain its ability to discriminate between a series of alkyd paints; of these packings, only the Carbowax 20M, Durapak, Porasil C, Porapak Q, and Chromosorb 102 had good discriminating power, though temperature programmes were used in each case. As used, the Carbowax 20M and Durapak have limited lifetimes. In this context, lifetime means the time in use before significant changes in retention times and/or resolution of peaks are noticeable. These packings were therefore rejected. We were unable to obtain column to column reproducibility with Porasil C, and so this packing was also rejected. Porapak Q and Chromosorb 102 gave similar results but Chromosorb 102 required a longer ageing period, and the batch used gave slightly inferior peak resolution compared with the Porapak Q. The latter was chosen as the column packing.

The following G.C. equipment and conditions were used:

Pyrolyser: Pye Unicam Curie-point, attached directly to G.C. column

Pyrolysis temperature: 610 °C, held for 10 seconds

G.C.: Pye 104, Model 64 (twin column, twin F.I.D.)

Column: Standard Pye 5 ft × 4 mm i.d. glass (silanised)

Carrier gas: Nitrogen

Carrier-gas flow rate: approximately 60 ml min^{-1}

Column packing: Porapak Q (50—80 mesh)

Temperature programme: 100—200 °C at 8 °C min^{-1} (held at 200 °C for up to 25 minutes; programmer started on completion of 10 s pyrolysis.

The columns were silanised prior to packing by passing a 5% solution of dichloro-dimethylsilane in toluene through the columns and were dried at 100 °C. Packing was facilitated by applying suction from the detector end of the column, accompanied by gentle tapping. The columns were aged by heating at 250 °C for 48 hours and with a nitrogen flow of 60 ml min^{-1}.

The hydrogen and air flows were adjusted to give approximately maximum senstivity. The flow rate required for pyrolysis was determined by injecting through the pyrolyser head 1 μl of headspace gas from a retention-time standard comprising 50 : 50 v/v methanol–n-propanol, with the column temperature at 100 °C, and immediately programming at 8 °C min^{-1}. The flow rate of nitrogen was adjusted to give retention times of 2.6 min and 9.1 min for methanol and n-propanol, respectively.

Having fixed the flow rate of carrier gas, the upper temperature limit was adjusted to give a retention time of 14.0 min for cyclohexane when injected at 100 °C, and the oven was programmed as described.

The pyrolysis wire was prepared in the following way. A portion (5—10 mm long) at one end of the wire was first flattened until the flattened portion was approximately 1 mm wide. A 'hook' was then made by doubling over about 2 mm of the flattened end. At this stage, the hook end plus a few centimetres of the wire near that end were heated to red heat in a bunsen flame to remove contaminants. The sample was placed in the elbow of the hook, care being taken not to touch or otherwise contaminate that end of the wire. The sample was then forced into close contact with the wire surface by crimping the hook. This method was used with all samples.

The mass spectral data were obtained from a linked P.G.C.–M.S. system. This consisted of a Pye Unicam Curie-point pyrolyser fitted to a Perkin-Elmer 880 G.C., which was linked *via* a Watson–Biemann-type diffusion separator to an AEI MS902 mass spectrometer.

G.C. conditions were basically the standard conditions described above, using Porapak Q (50—80 mesh) packed in 5 ft glass columns, as with the Pye 104 equipment. In order to obtain compatibility with the mass spectrometer, the following exceptions to the standard conditions were made. Columns were of i.d. 1/8 in and the carrier gas was helium, with a flow rate of 20 ml min.$^{-1}$ The chromatograms were monitored by recording the total ion current in the mass spectrometer, which was operated at 70 eV. Sample sizes of several hundred micrograms were used in order to obtain an intense spectrum for all the peaks in the pyrogram.

The mass spectrum for each peak was scanned near or at the peak maximum, and wherever possible a background spectrum was run between the peaks for comparison. This was particularly important when the temperature programme was nearing its end, when the background bleed became more intense. Most of the substances thus identified were injected under standard conditions to check that their retention times matched those in the pyrograms.

Results and Discussion

The pyrograms from 43 polymers and polymer mixtures are presented here (copies of the original pyrograms plus those of 22 other substances falling under our heading of plastics and paints are available on request). The polymers were not pretreated in any way, and they were pyrolysed in the solid phase as received.

The two main criteria are (a) the reproducibility of the system and (b) the effectiveness of the method as a means of discriminating between polymers.

(a) *Reproducibility*. An important reason for choosing the polymer bead packing Porapak Q is that is can be used without preparation and has a long life under the conditions used in this work. Reproducibility from column to column is thus mainly dependent on the manufacturer's quality control.

Although the work was restricted to the instrumental combination of Pye Unicam Curie-point pyrolyser and Pye 104 Model 64 chromatograph, three completely independent systems of this type were used. Possible sources of instrument error are the mass-flow controllers, the temperature programmers, and the temperature of the column oven. The mass-flow controllers were found to be acceptable, though checks of the retention times of a mixture of methanol, n-propanol, and cyclohexane should be made at least daily. The time of mechanical rotation of the programmer dials from 100 to 200°C at 8°C min^{-1} varied by ± 6 seconds (the mean was 12 minutes 30 seconds). This is equivalent to ± 0.8°C. This variation was probably due to the difficulty in accurately fixing the initial and final temperature pegs, rather than in differences in programming rates. The temperature measured by a mercury-in-glass thermometer continued to rise after the end of the programme, rising 2°C in the first 3 minutes of the final period and a further 1°C in the following 7 minutes (at 200°C). This error was constant on the three chromatographs.

The three P.G.C. instruments were each used with the same analysing column, and 10 pyrograms were produced from a single alkyd paint on each instrument. The pyrograms contain the same peaks, seen in pyrograms A and B (Figure 5.1). The first three peaks have retention times of less than one minute and are of unreliable relative peak height. The heights of peaks a—j were measured and the individual heights compared to the sum of the heights of these peaks.

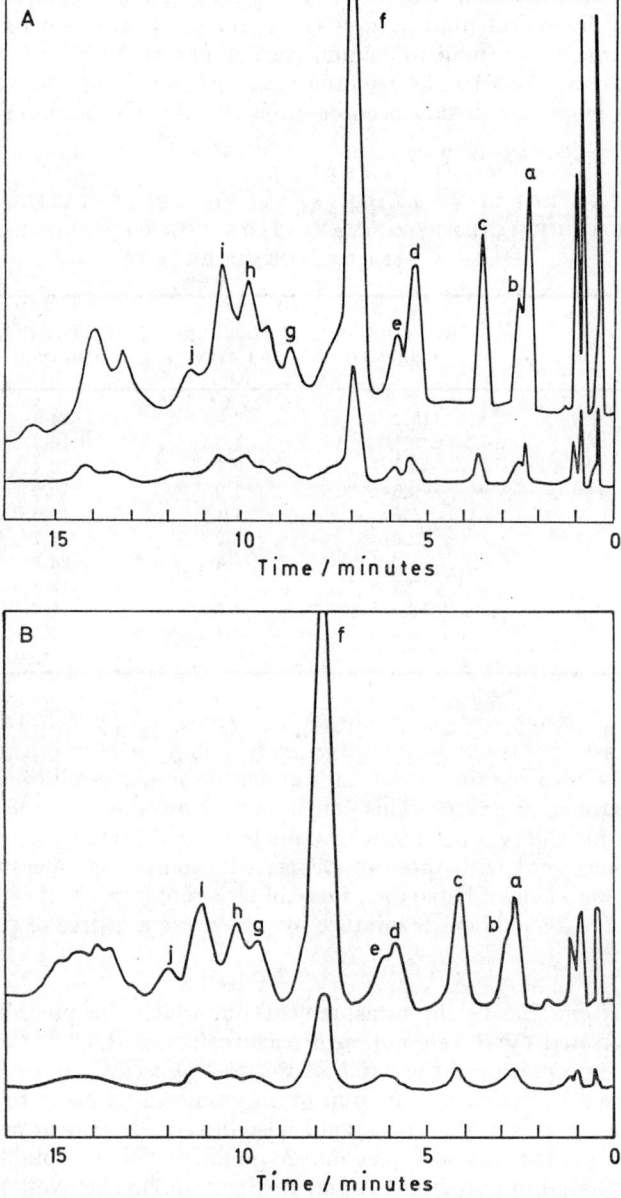

Fig. 5.1 Acceptable limits of resolution in pyrograms from an alkyd paint. Upper trace, f.s.d. with 10 mV input; lower trace, f.s.d. with 50 mV input

Table 5.1 shows the mean relative peak heights and their coefficients of variation due to differences from pyrolysis to pyrolysis, instrument to instrument (three instruments), column to column (two columns), and batch to batch of packing material (four batches) in the standard deviations derived from 90 results. This work has already been described by the authors in more detail.[19]

TABLE 5.I

MEAN RELATIVE HEIGHTS, STANDARD DEVIATIONS,
AND COEFFICIENTS OF VARIATION FOR 90 PYROGRAMS
OF A SINGLE ALKYD PAINT

Peak	Mean relative height	Standard deviation	Coefficient of variation
a	13.7	2.0	14.6
b	8.4	1.4	16.3
c	6.6	0.7	10.4
d	9.6	1.0	10.0
e	5.6	1.1	19.6
f	22.9	2.2	18.7
g	5.6	1.4	24.8
h	7.7	1.1	13.8
i	13.1	1.3	9.7
j	6.4	1.4	21.5

In the case of complex pyrograms such as those from which Table 5.1 was derived, measurements of peak height were found to be more precise than those of peak area obtained from an automatic base-line-following shoulder-sensing electronic integrator. This can be seen by comparing the coefficients of variation for the two peak 'size' parameters tabulated in Table 5.2. These results are in accord with those of Deans, who found that measurements of peak height were more reliable than those of integrated area in the quantitative estimation of a six-component mixture by g.c. where a degree of peak overlap was present.[20]

(i) Resolution. The height measurements on which the preceding work is based are affected by the resolution of each column, which is largely determined by the properties of the batch of Porapak Q used. The recognition of a pattern of peaks forming a pyrogram of an unknown as being that found in one of the pyrograms in this collection is also heavily dependent on the degree of resolution of the analyser column. A guide to the acceptable limits of resolution is found in pyrograms A and B (Figure 5.1). The overlapping peaks (a) and (b) (propylene and propane, respectively) which elute immediately before the discrete acetaldehyde peak (c) and the overlapping peaks for butene and butane [(d) and (e), respectively] which appear between acetaldehyde and

acrolein (f) are used as resolution standards. The degree of overlap of these two pairs of peaks should fall between that found in pyrogram A and that in pyrogram B, judged by visual inspection, when a gloss household alkyd paint is pyrolysed. A column with resolution outside these limits may give pyrograms with patterns which it will be difficult or impossible to compare successfully with this collection. It is recommended that 50—80 mesh Porapak Q be used, but if the resolution proves poorer than that in pyrogram B then the use of a finer mesh size, such as 80—100 mesh, will improve it. The C_3 and C_4 alkene-alkane pairs seen in these two pyrograms are common to other polymers with

TABLE 5.2

COMPARISON OF THE PRECISION OF MEASUREMENTS OF RELATIVE HEIGHT AND RELATIVE AREA IN PYROGRAMS WHERE MOST PEAKS ARE INCOMPLETELY RESOLVED

Peak	Mean relative height	Coefficient of variation (%)	Mean relative area	Coefficient of variation (%)	Comments
a	14.2	2.0	6.6	19.1	} Overlapping
b	8.4	1.9	3.3	10.0	} peaks
c	6.5	1.8	3.6	11.6	
d	9.5	0.9	8.0	6.7	} Overlapping
e	5.1	3.5	3.6	10.5	} peaks
f	22.8	6.5	21.6	3.5	
g	4.8	2.9	5.2	15.0	⎫
h	7.5	8.1	9.2	8.1	⎬ Overlapping
i	12.8	5.2	15.9	4.6	⎱ peaks
j	7.8	11.4	22.6	12.4	⎭

long hydrocarbon chains. However, the use of a gloss household alkyd paint is essential in testing the resolution of a column compared with pyrograms A and B, as the relative proportions of these C_3 and C_4 peaks will differ in other types of polymer.

The pyrograms A and B form a visual representation of the variations in resolution which lead to differences in relative peak heights. The resultant slight variations in the appearance of the pyrograms have to be considered when using this reference collection (see following section on the index) but it must be remembered that better discrimination is possible when one column is used to produce two pyrograms for comparison. Thus, although two different polymers may be identified only as being of the same general class by using this reference collection, it is often possible to distinguish between them by consecutive pyrolyses of the polymers on one column.

(b) *Discriminating Power.* All the polymers investigated gave a pyrogram consisting of at least one peak, but each peak is not necessarily a distinguishing characteristic of a single polymer. This is especially so of the peaks of very low retention time, for example those from the hydrocarbons of low molecular weight which are evolved by the pyrolysis of many polymers. When only one significant peak is present in a pyrogram it frequently represents a monomer. In these cases the reliability of identification is clearly related to that attained when a volatile compound is identified from one retention time on one column packing; however, because only a limited group of polymers pyrolyse entirely to monomer, this is a distinguishing characteristic in itself.

The discriminating power (and sensitivity) of the method is dependent on the type of detector as well as the column. Some polymers with a high hetero-atom content yield compounds which cause little or no response in a F.I.D. These products are usually very small molecules, *e.g.* the nitrogen oxides, and consequently are of limited diagnostic value.

The ability of the method to discriminate between polymers is demonstrated in the collection of pyrograms. The pyrogram patterns are not completely independent of the weight of polymer pyrolysed, but this weight dependence is reduced and becomes unimportant in most cases when small quantities are pyrolysed; 5—$20\,\mu g$ is a useful range of weights, giving good reproducibility of pyrograms (a low concentration of pyrolysate vapour in the pyrolyser reduces the chance of secondary reactions). A much wider range of weights can be used with polymers which pyrolyse by a process of chain depolymerisation from one end (unzipping), giving high yields of monomers.

(c) *Sensitivity.* With the G.C. operating conditions, detector, amplifier attenuation, column packing, and column length fixed, the sensitivity of the method to the analysis of each polymer is indicated by the weight of polymer required to produce each pyrogram (see the library of pyrograms). The sensitivity of the method to individual monomers in a copolymer is dependent on the structure of the polymer. Small percentages of a monomer can be undetectable if incorporated in an alternating polymer. A possible reason for this can be seen in Figure 5.2, which represents a length of copolymer chain formed by monomers A and B. (1) represents a section of random copolymer, (2) a section of alternating copolymer, (3) a section of block copolymer. If the homopolymers of A and B pyrolyse by 'unzipping', and if the A–B bond is strong compared to the A–A and B–B ones, it is probable that only the underlined monomer units will form free monomer on pyrolysis.

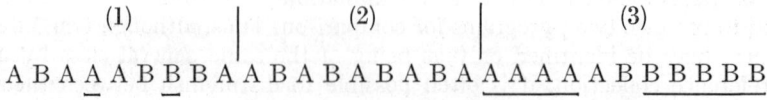

(1)   (2)   (3)

A B A A A B B B A A B A B A B A B A A A A A A B B B B B B

Fig. 5.2   Representation of a polymer formed of monomers A and B.

(*d*) *Index System*. As explained earlier in this chapter, a visual comparison of the pyrogram of an unknown with the pyrograms in this collection will be required to establish the identity of the unknown. An index system has been derived in order to reduce the number of comparisons required. The collection is too small to warrant the use of a more complex indexing and retrieval system at this stage.[21]

(*i*) *Use of the index*. The retention times of the three highest peaks in the pyrogram of an unknown are noted in order of increasing retention time, *not* peak height. (Retention times are measured from the end of the 10 second pyrolysis period.) Heavily overlapping peaks such as the butane–butene ones are considered as one peak, the retention time being taken from the higher of the two.

Exceptions:
1. Peaks of retention times less than 1.5 minutes are ignored.
2. To avoid confusion where several small peaks of similar height are present, the second- and third-highest peaks are considered only if the peak height is at least 10% of the highest peak.

The one or more rows in the index (Tables 5.3 and 5.4) into which the three (or less) retention times can be fitted give the pyrograms which should be referred to for comparison to determine the nature of the unknown.

Variations of peak height of up to 30% have been allowed for in producing the index. This figure is approximately twice the mean of the coefficients of variation for the ten peaks obtained from 90 pyrograms of an alkyd paint shown in Table 5.1. It is unlikely that errors in the measurement of the relative height of the three highest peaks of any pyrogram will be greater than those for the ten peaks of Table 5.1 since the latter exhibit a much wider range of relative peak height and include several heavily overlapping peaks. Large margins of error in retention times, $\pm 10\%$ for retention times greater than five minutes and $\pm 15\%$ for retention times less than five minutes, have been allowed for in both the reference and unknown pyrograms. These tolerances in peak height and retention time are incorporated in the index, so that only absolute values are required from the pyrograms of an unknown if one is to make successful use of the index. The index is in two parts, plastics (Table 5.3) and paint polymers (Table 5.4); each table follows the relevant section of the pyrogram library. The pyrogram library is being extended into other fields of forensic interest, such as the identification of synthetic fibres and rubber.

The Pyrogram Library

The library is divided into two sections: (*a*) plastics, (*b*) paint polymers. All pyrograms were obtained with an attenuation setting of 50 on the ionisation amplifier. A two-pen recorder was used, the upper trace showing full-scale deflection with a 10mV input, the lower trace showing full-scale deflection at 50mV input. The original pyrograms were obtained with a recorder chart speed of two centimetres per minute.

Plastics (Pyrograms 1–29)

This section contains pyrograms of polymers used in applications where some degree of structural strength is required, but also includes polymers used in film and tape form. Some of the polymers are also found in paints, causing overlap between the sections.

Table 5.5 provides a key to the numbered peaks which were identified or had their identities confirmed from their mass spectra.

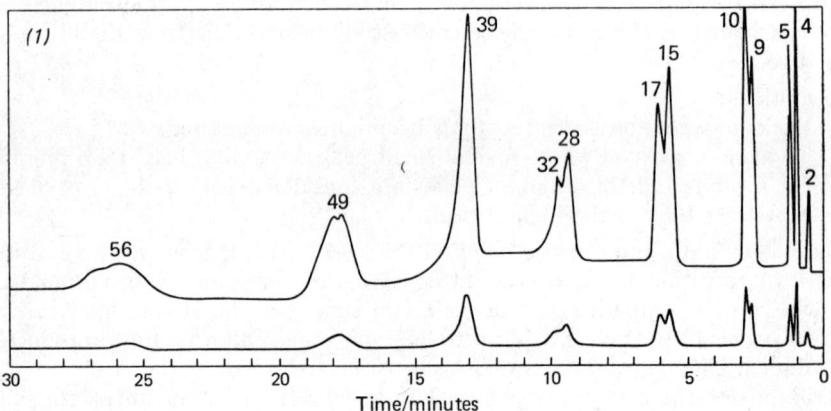

Pyrogram 1: Polyethylene (low density, high pressure). High density polyethylene gives a similar pyrogram   Amount pyrolysed: 11 μg

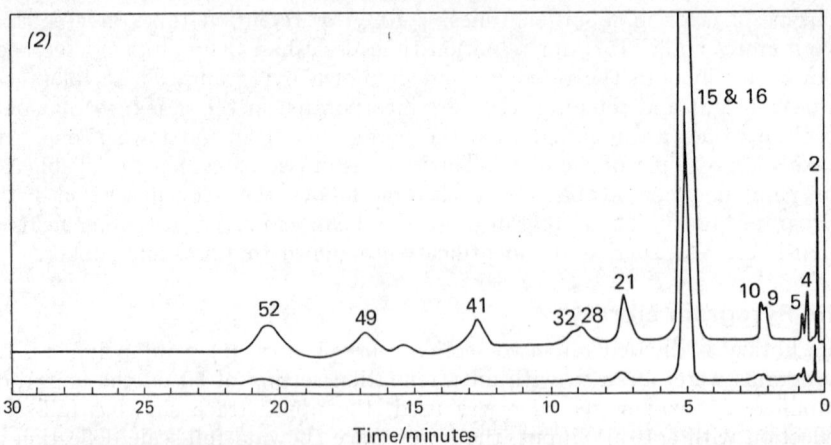

Pyrogram 2: Copolymer of ethylene (45%) and isobutylene (55%)

Amount pyrolysed: 2 μg

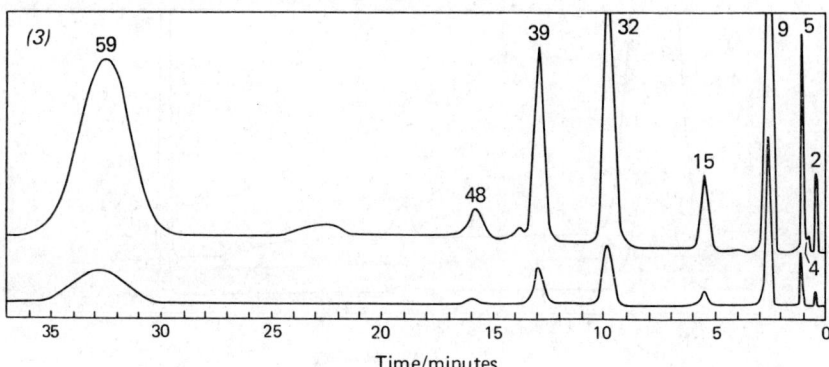

Pyrogram 3: Polypropylene   Amount pyrolysed: 5 μg

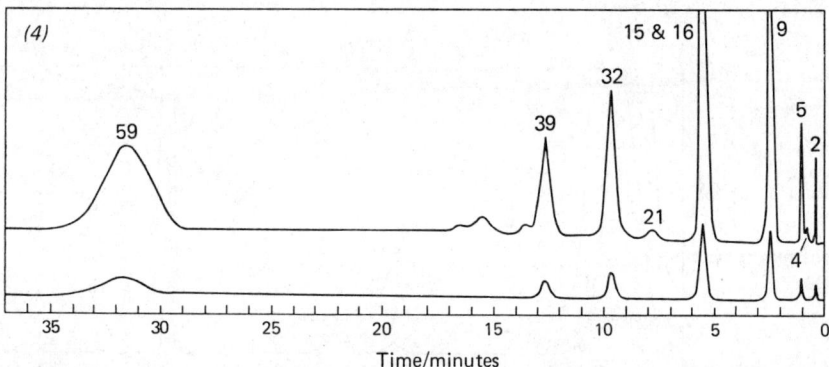

Pyrogram 4: Copolymer of propylene (90%) and isobutylene (10%)
                                    Amount pyrolysed: 3 μg

Pyrogram 5

Copolymer of
        butadiene
   (68%) and
     acrylonitrile
   (32%)

Amount pyrolysed:
   7 μg

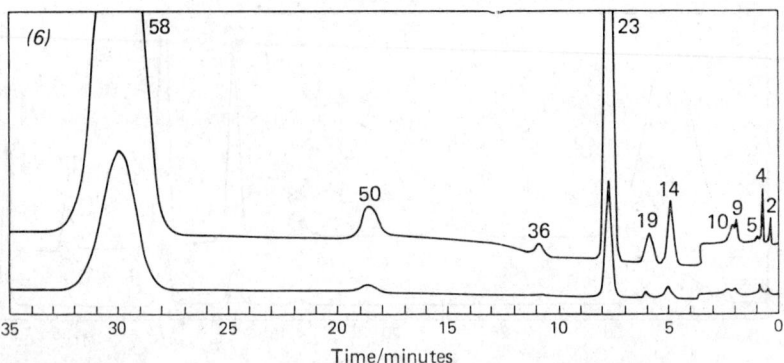

Pyrogram 6: Copolymer of styrene (64%), acrylonitrile (26%), and butadiene (10%)   Amount pyrolysed: 25 μg

Pyrogram 7

Copolymer of styrene (75%) and acrylonitrile (25%)

Amount pyrolysed: 49 μg

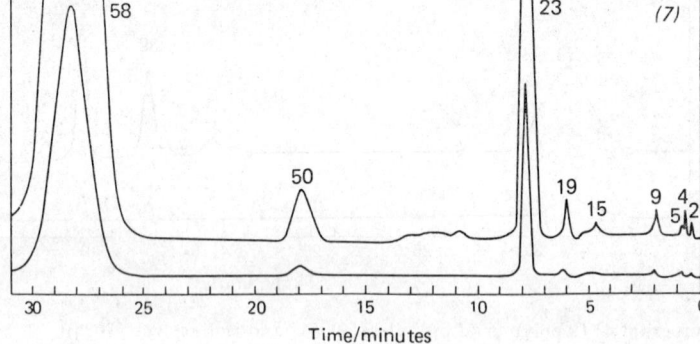

Pyrogram 8

Styrene–butadiene rubber

Amount pyrolysed: 15 μg

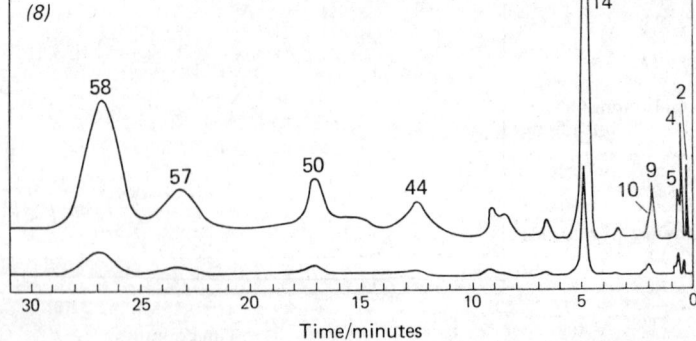

Pyrogram 9

Soft vulcanised natural
   rubber

Amount pyrolysed: 6 μg

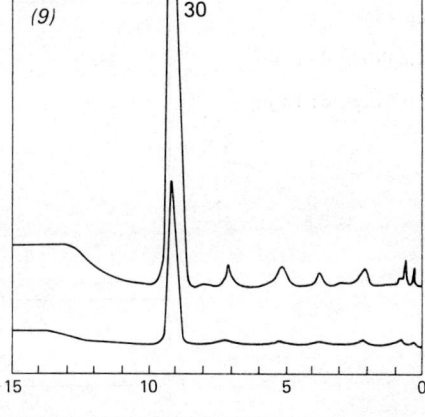

Pyrogram 10

Polytetrafluoroethylene

Amount pyrolysed: 8 μg

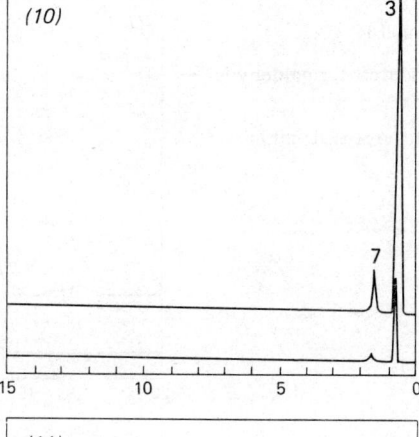

Pyrogram 11

Polyformaldehyde

Amount pyrolysed: 8 μg

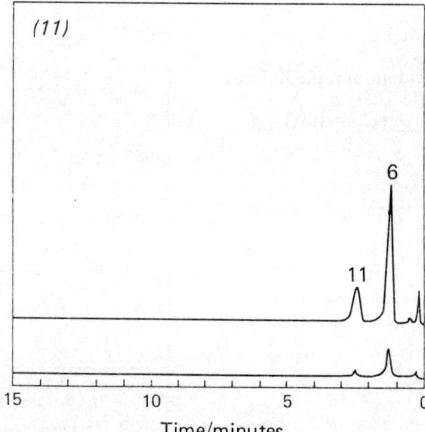

Time/minutes

Pyrogram 12

Urea–formaldehyde resin

Amount pyrolysed: 15 μg

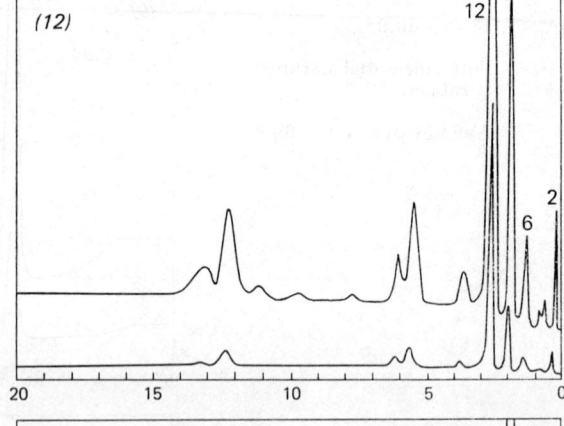

Pyrogram 13

Urea/thiourea–formaldehyde
    resin

Amount pyrolysed: 20 μg

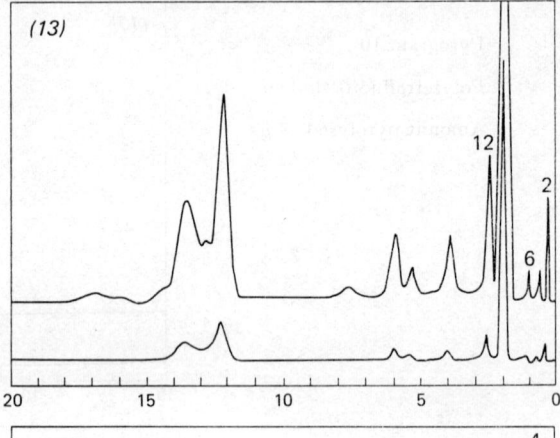

Pyrogram 14

Poly(ethylene terephthalate)

Amount pyrolysed: 11 μg

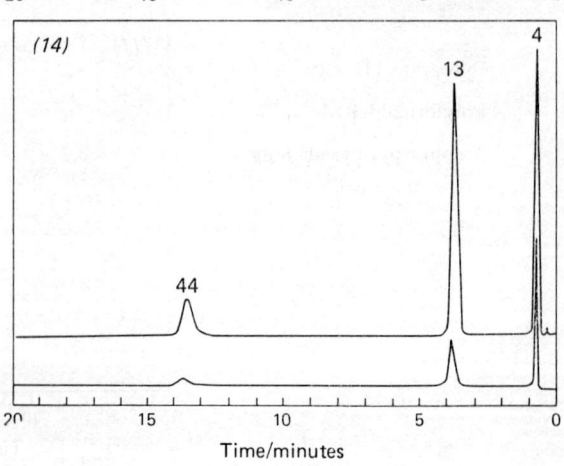

Time/minutes

Pyrogram 15

Poly(methyl acrylate)

Amount pyrolysed: 4 μg

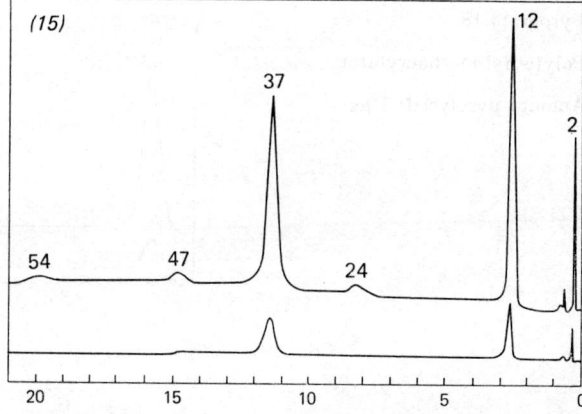

Pyrogram 16

Poly(ethyl acrylate)

Amount pyrolysed: 1 μg

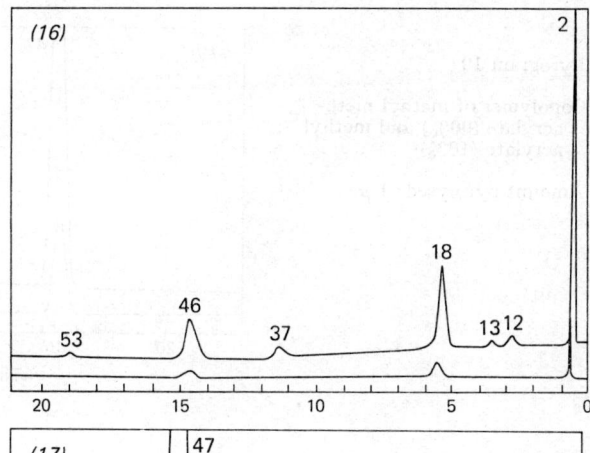

Pyrogram 17

Poly(methyl methacrylate)

Amount pyrolysed: 1 μg

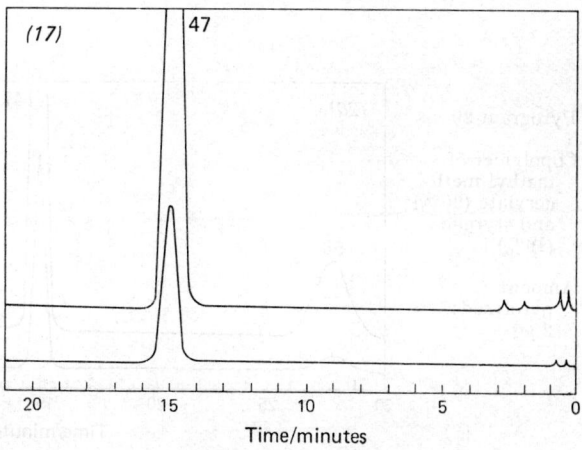

Time/minutes

Pyrogram 18

Poly(ethyl methacrylate)

Amount pyrolysed: 1 μg

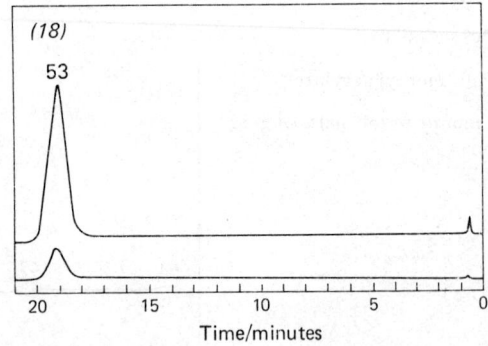

Pyrogram **19**

Copolymer of methyl meth-
  acrylate (90%) and methyl
  acrylate (10%)

Amount pyrolysed: 1 μg

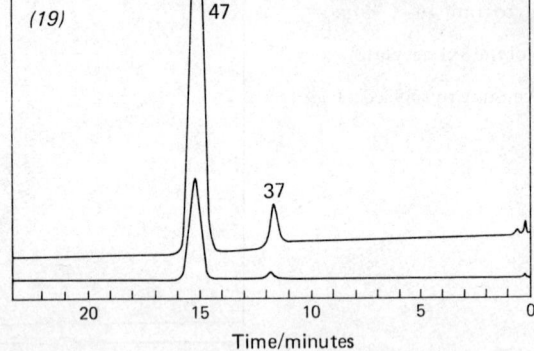

Pyrogram 20

Copolymer of
  methyl meth-
  acrylate (90%)
  and styrene
  (10%)

Amount
  pyrolysed:
  2 μg

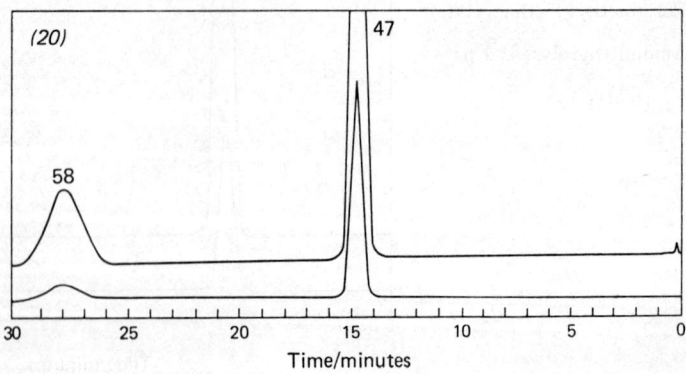

Pyrogram 21

Copolymer of vinyl alcohol
  (80%) and vinyl acetate
  (20%)

Amount pyrolysed: 20 μg

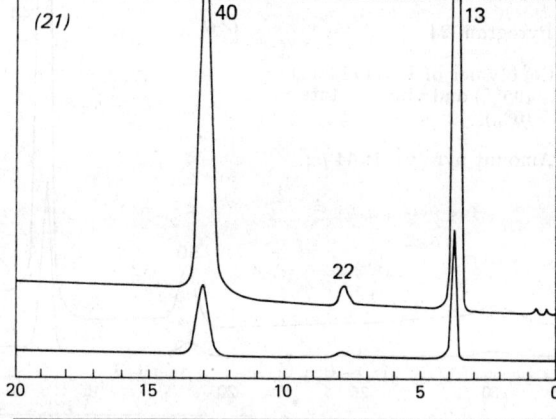

Pyrogram 22

Poly(vinyl acetate)

Amount pyrolysed: 2 μg

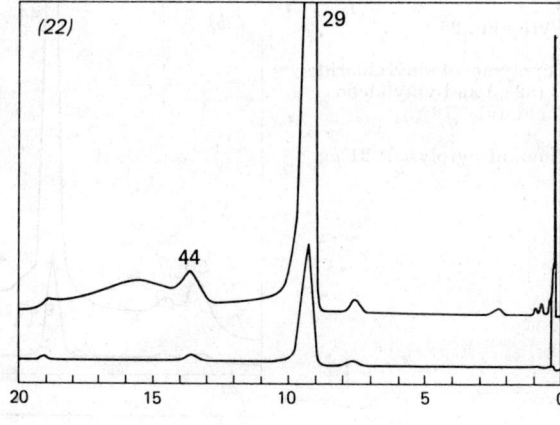

Pyrogram 23

Poly(vinyl chloride)

Amount pyrolysed: 12 μg

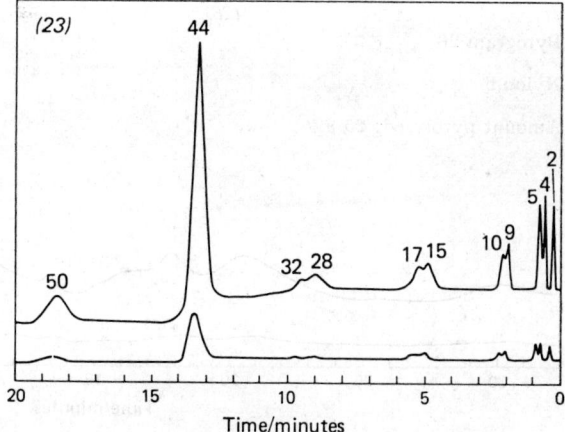

Time/minutes

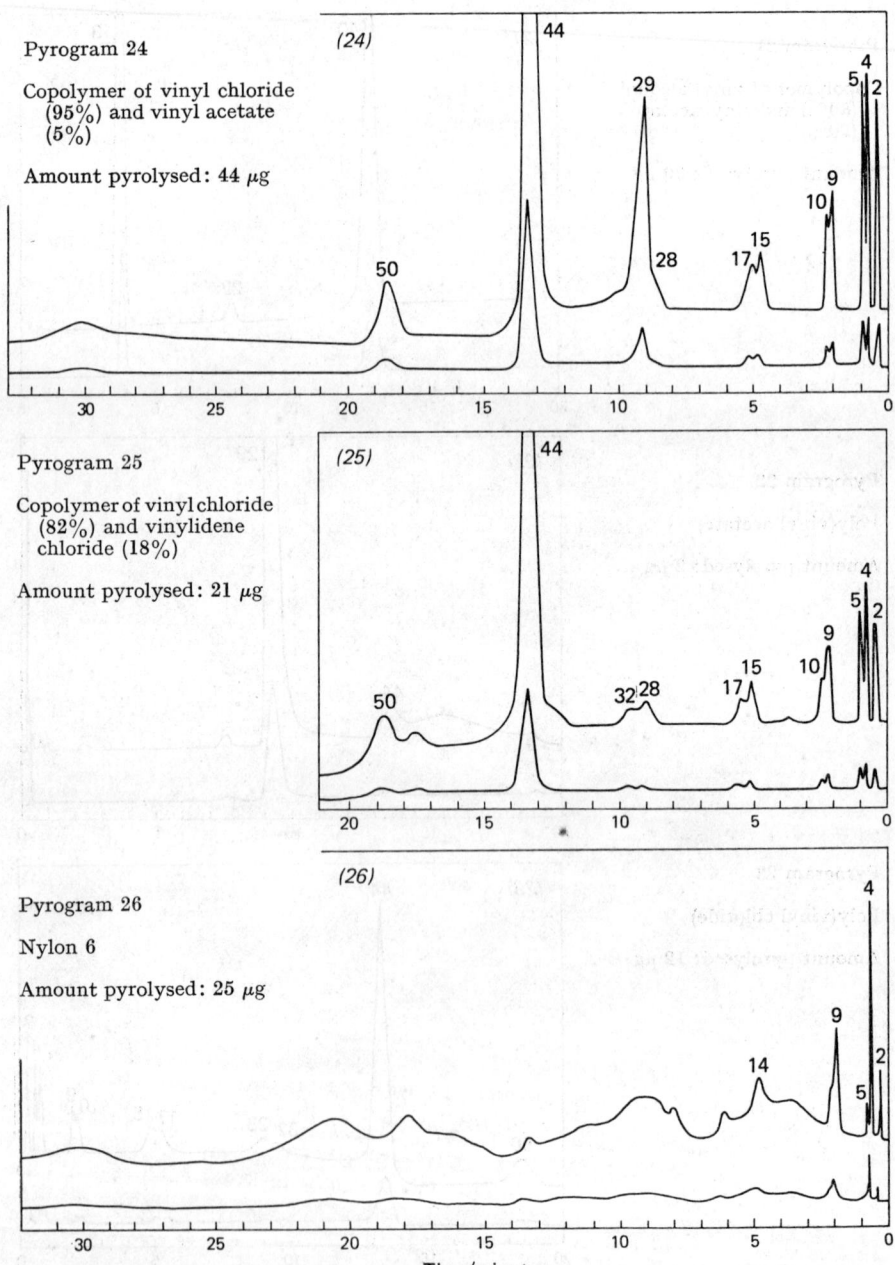

Pyrogram 24

Copolymer of vinyl chloride
   (95%) and vinyl acetate
   (5%)

Amount pyrolysed: 44 μg

Pyrogram 25

Copolymer of vinyl chloride
   (82%) and vinylidene
   chloride (18%)

Amount pyrolysed: 21 μg

Pyrogram 26

Nylon 6

Amount pyrolysed: 25 μg

Time/minutes

Pyrogram 27

Nylon 66

Amount pyrolysed: 8 µg

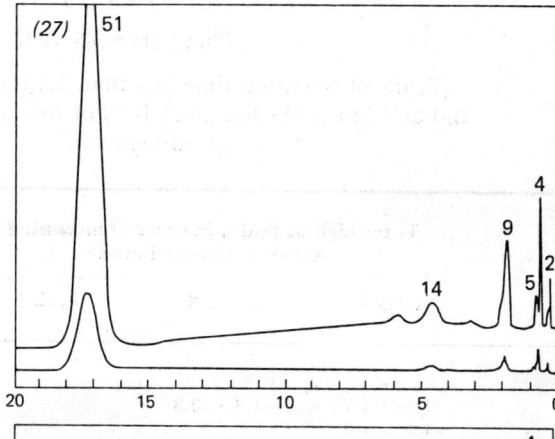

Pyrogram 28

Nylon 610

Amount pyrolysed: 12 µg

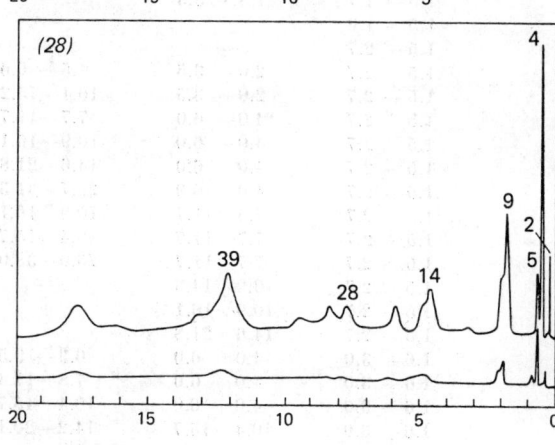

Pyrogram 29

Nylon 66/610 (40:30)

Amount pyrolysed: 12 µg

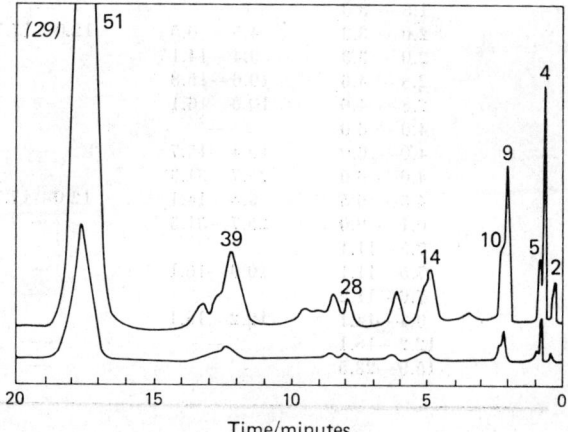

Time/minutes

TABLE 5.3

PLASTICS INDEX

(Peaks of retention time less than 1.5 minutes are ignored;
2nd and 3rd peaks less than 10% of the height of the 1st peak
are ignored)

Three highest peaks, in order of increasing retention times/minutes			Corresponding pyrograms
Peak 1	Peak 2	Peak 3	
—	—	—	10
1.5— 1.7	1.8— 3.3	—	11
1.5— 1.9	—	—	10, 26
1.5— 2.7	—	—	10, 26
1.5— 2.7	2.0— 3.5	4.5— 6.6	12
1.5— 2.7	2.0— 3.5	10.1—15.2	12
1.5— 2.7	4.0— 6.0	7.7—11.7	4, 5
1.5— 2.7	4.0— 6.0	10.9—16.1	4, 23
1.5— 2.7	4.0— 6.0	14.6—21.8	27
1.5— 2.7	4.0— 6.0	25.7—31.3	8
1.5— 2.7	7.5—11.1	10.9—16.1	3, 24
1.5— 2.7	7.7—11.7	10.4—15.7	3, 24
1.5— 2.7	7.7—11.7	26.0—38.0	3
1.5— 2.7	10.0—14.8	—	13, 23
1.5— 2.7	10.9—16.1	—	23
1.5— 2.7	14.6—21.8	—	27
1.6— 3.0	4.0— 6.0	6.3— 9.5	4, 5
1.6— 3.0	4.0— 6.0	7.8—11.4	4, 5
1.6— 3.0	4.0— 6.0	10.4—15.7	1, 4, 16
1.6— 3.0	10.4—15.7	14.2—20.1	1, 23
1.8— 3.3	—	—	11
2.0— 3.3	4.5— 6.5	12.0—17.7	1, 16
2.0— 3.3	9.4—14.1	—	15, 23
2.8— 4.6	10.6—15.8	—	14, 21
2.8— 4.6	10.9—16.1	—	14, 21
4.0— 6.0	—	—	2
4.0— 6.0	10.4—15.7	—	14, 21
4.0— 6.0	25.7—31.3	—	8
4.5— 6.5	9.4—14.1	12.0—17.7	16
6.1— 9.0	25.7—31.3	—	6, 7
7.5—11.1	—	—	9, 22
7.5—11.1	10.9—16.1	—	19, 22
7.6—11.2	—	—	9, 22
9.4—14.1	12.2—18.1	—	19
12.2—18.1	—	—	17, 19, 20
15.9—23.5	—	—	18, 19, 20

Paint Polymers (Pyrograms 30—43)

This section includes a selection of pigmented and unpigmented polymers commonly used as paint vehicles. Examination of specially prepared resins and commercial paints of the alkyd type containing a variety of inorganic and organic pigments has shown that, on pyrolysis, the pigments did not interfere with the recognition of the pyrolysis patterns, with the possible exception of carbon. It was believed to be because of interference from carbon that some of the commercial black paints, although easily seen to be alkyds from their pyrograms, gave a significantly different pyrolysis pattern compared with paints of a similar type but of different colour. The most marked difference was an almost complete suppression of the production of acrolein; an attempt was made to simulate a black paint by mixing carbon black with a dry resin, and this mixture produced a partial suppression of acrolein formation on pyrolysis compared with the unpigmented resin. The less marked effect of the pigment in this case may be due to the less intimate mixing of the two components than is normally the case in a commercial paint; however, it is also possible that resins which are inferior or different in some respect are used in black paints. The only other effect attributable to the presence of pigments was the occasional appearance of a peak immediately prior to the propylene peak in the alkyds; such behaviour was associated with some organic pigments.

Pyrogram 30

Linseed oil glycerol
  o-phthalate alkyd

Amount pyrolysed: 15 µg
  of pigmented paint

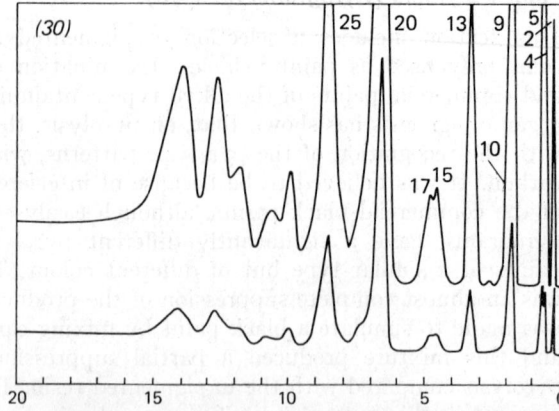

Pyrogram 31

Hydrogenated castor oil
  glycerol o-phthalate alkyd

Amount pyrolysed: 10 µg
  of unpigmented resin

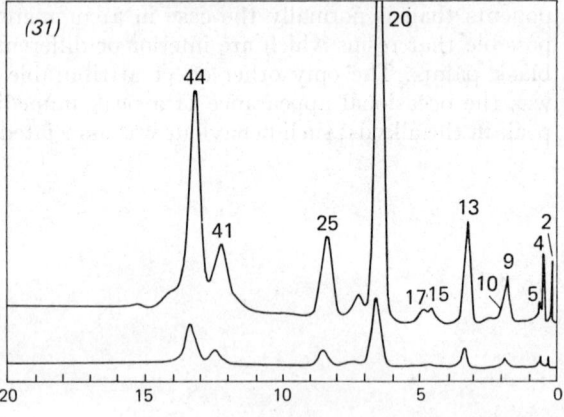

Pyrogram 32

Coconut oil glycerol glycol
  o-phthalate alkyd

Amount pyrolysed: 14 µg
  of unpigmented resin

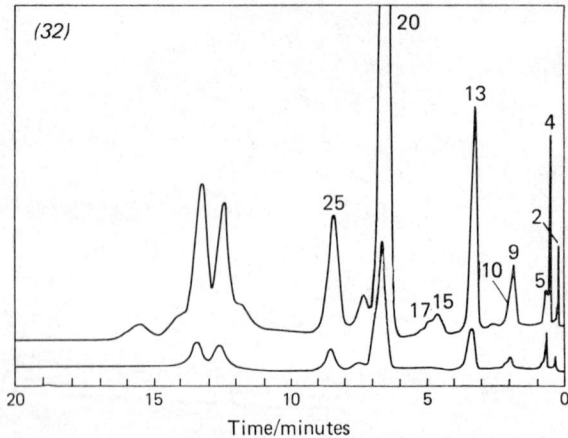

Time/minutes

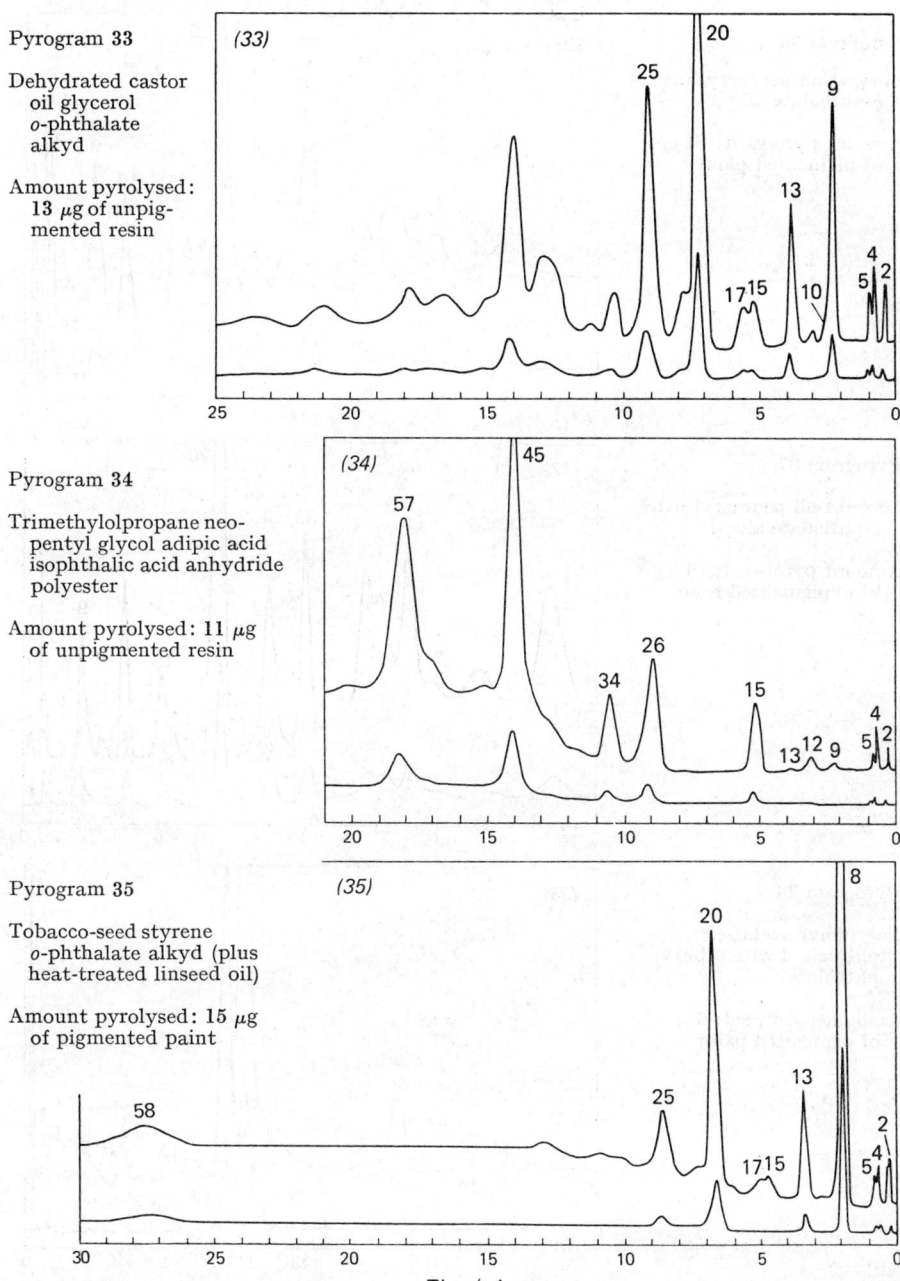

Pyrogram 33

Dehydrated castor oil glycerol o-phthalate alkyd

Amount pyrolysed: 13 μg of unpigmented resin

Pyrogram 34

Trimethylolpropane neopentyl glycol adipic acid isophthalic acid anhydride polyester

Amount pyrolysed: 11 μg of unpigmented resin

Pyrogram 35

Tobacco-seed styrene o-phthalate alkyd (plus heat-treated linseed oil)

Amount pyrolysed: 15 μg of pigmented paint

Time/minutes

Pyrogram 36

Linseed oil pentaerythritol
  o-phthalate alkyd

Amount pyrolysed: 14 µg
  of pigmented paint

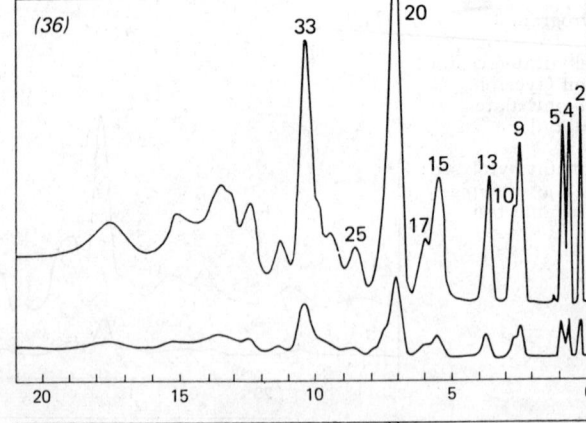

Pyrogram 37

Coconut oil pentaerythritol
  o-phthalate alkyd

Amount pyrolysed: 16 µg
  of unpigmented resin

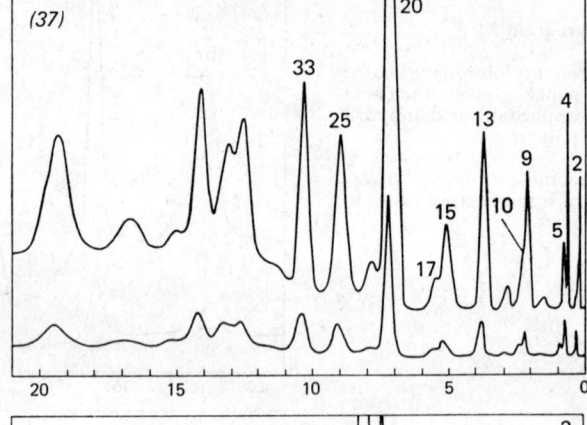

Pyrogram 38

Poly(vinyl acetate),
  plasticised with dibutyl
  phthalate

Amount pyrolysed: 6 µg
  of pigmented paint

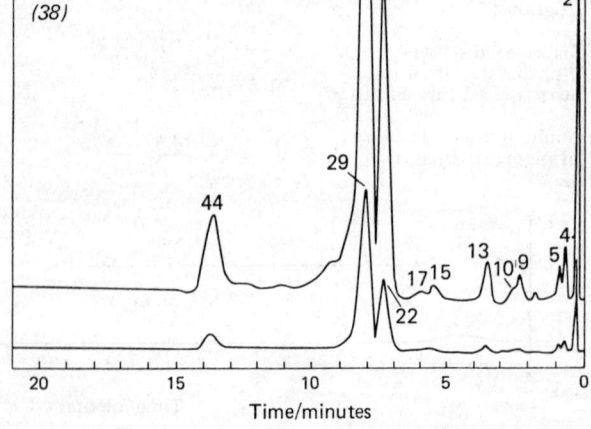

Time/minutes

Pyrogram 39

Copolymer of vinyl acetate (85%) and vinyl caprate (15%)

Amount pyrolysed: 8 μg of pigmented paint

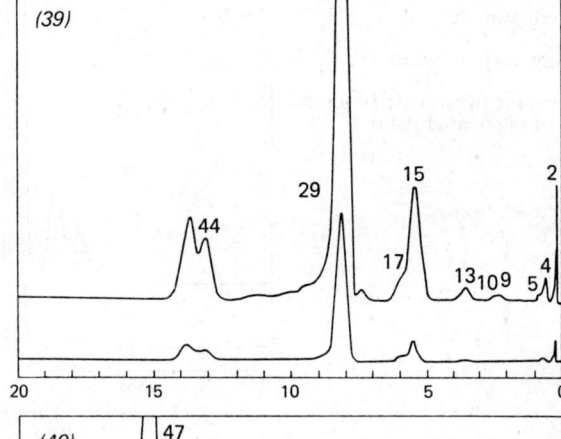

Pyrogram 40

Acrylic vehicle enamel (solvent type)

Amount pyrolysed: 7 μg

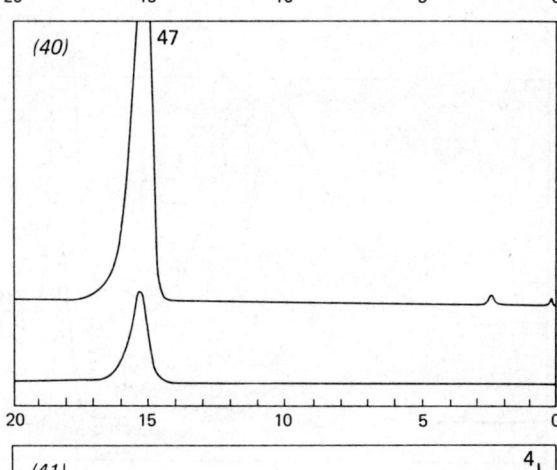

Pyrogram 41

Acrylic emulsion

Amount pyrolysed: 7 μg

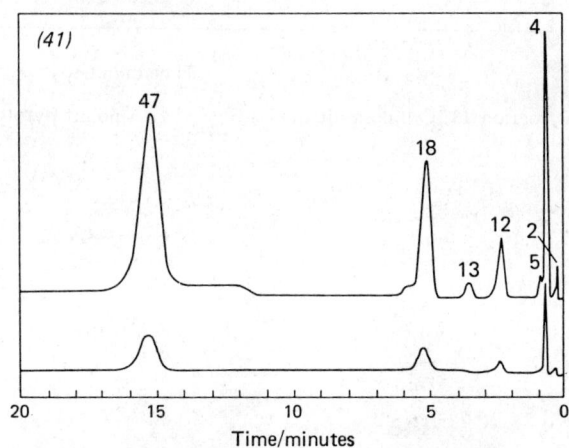

Time/minutes

Pyrogram **42**

Poly(vinyl butyral)

Amount pyrolysed: **10** μg of pigmented paint

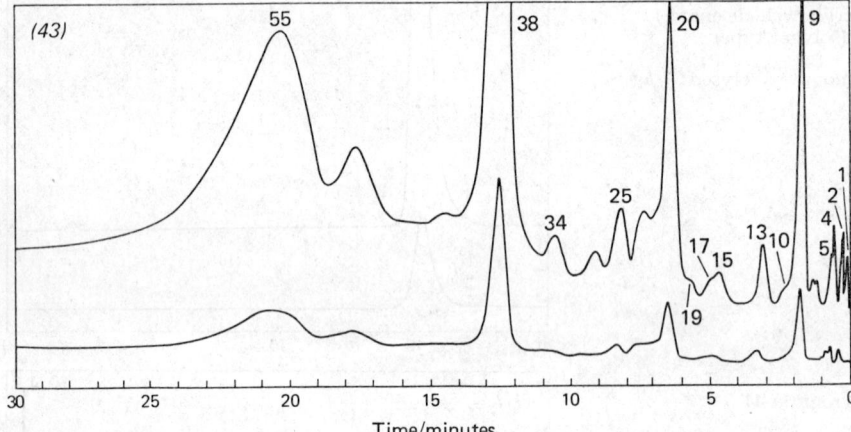

Pyrogram **43**: Cellulose nitrate   Amount pyrolysed: 8 μg of pigmented paint

TABLE 5.4

PAINT POLYMER INDEX

(Peaks of retention time less than 1.5 minutes are ignored; 2nd and 3rd peaks less than 10% of the height of the 1st peak are ignored. The pyrogram of an alkyd can be recognised by the series of peaks 2, 4, 5, 9, 10, 13, 15, 17, 20, 25, 33 of pyrograms 30—37; where peak 33 is larger than peak 25 the alkyd contains pentaerythritol rather than glycerol as the polyhydric alcohol; where peak 25 is large and peak 33 absent or small, the reverse is true. Apart from pyrograms 34 and 35, which are not easily recognised as alkyds, the alkyd pyrograms 30—37 are not included in this index.)

Three highest peaks, in order of increasing retention times/minutes			Corresponding pyrograms
Peak 1	Peak 2	Peak 3	
1.3— 2.5	2.7— 4.2	5.5— 9.2	35
1.3— 2.5	5.5— 8.2	—	35
1.5— 2.7	5.5— 8.2	10.1—14.9	41, 43
1.9— 3.4	4.4— 6.5	12.2—18.1	41, 43
2.7— 4.2	8.6—12.8	9.6—14.4	42
4.0— 6.0	7.3—10.9	—	39
4.0— 6.0	7.3—10.9	10.8—16.2	39
6.2— 9.3	7.3—10.9	—	38
7.2—10.8	11.2—16.8	14.4—22.1	34
8.4—12.8	9.6—14.4	—	42
12.2—18.1	—	—	40

TABLE 5.5

KEY TO KNOWN PYROLYSIS PRODUCTS IN ORDER OF INCREASING RETENTION TIMES

Peak No.	Compound	Retention time/ minutes	Peak No.	Compound	Retention time/ minutes
1	Carbon monoxide	<1.5	8	Hydrogen cyanide	2.0
2	Methane	<1.5	9	Propylene	2.1
3	Tetrafluoroethylene	<1.5	10	Propane	2.3
4	Ethylene	<1.5	11	Dimethyl ether	2.5
5	Ethane	<1.5	12	Methanol	2.6
6	Formaldehyde	<1.5	13	Acetaldehyde	3.5
7	Hexafluoropropylene	1.5	14	Buta-1,3-diene	5.0

TABLE 5.5 (*continued*)

Peak No.	Compound	Retention time/ minutes	Peak No.	Compound	Retention time/ minutes
15	n-Butene	5.0	39	Hexene	12.8
16	Isobutene	5.1	40	Crotonaldehyde	13.0
17	n-Butane	5.4	41	Hexane	13.0
18	Ethanol	5.5	42	3-Methylhexane	13.2
19	Acetonitrile	5.8	43	Hexa-2,4-diene	13.2
20	Acrolein	6.8	44	Benzene	13.4
21	Neopentane	7.3	45	Tiglaldehyde	14.0
22	Acetone	7.7	46	Ethyl acrylate	14.7
23	Acrylonitrile	7.8	47	Methyl methacrylate	15.0
24	Methyl acetate	8.3	48	2,4-Dimethylpent-1-ene	15.5
25	Allyl alcohol	8.5			
26	2-Methylbutene	8.7	49	Heptenes (mixture)	17.5
27	Penta-1,3-diene	8.8	50	Toluene	18.3
28	n-Pentene	9.1	51	Cyclopentanone	18.3
29	Acetic acid	9.1	52	Octane	19.5
30	Isoprene	9.3	53	Ethyl methacrylate	19.5
31	Propionitrile	9.5	54	Methyl ethacrylate	20.0
32	n-Pentane	9.6	55	2-Isopropoxyethanol	20.5
33	Methacrolein	10.3	56	Octenes (mixture)	22.3
34	Isobutyraldehyde	10.4	57	4-Vinylcyclohex-1-ene	24.5
35	n-Butyraldehyde	10.6	58	Styrene	28.5
36	Crotononitrile	10.9	59	2,4-Dimethylhept-1-ene	32.0
37	Methyl acrylate	11.6			
38	n-Butanol	12.4			

References for Chapter 5

1 Evans, M. B., and Smith, J. F., *J. Chromatog.*, 1961, **6**, 293.
2 Leathard, D. A., and Shurlock, B. C., 'Identification Techniques in Gas Chromatography,' Wiley–Interscience, London and New York, 1970, p. 58.
3 Kovats, E., *Helv. Chim. Acta*, 1958, **41**, 1915.
4 Kovats, E., *Adv. Chromatog.*, 1966, **1**, 229.
5 Widmer, H., *J. Gas Chromatog.*, 1967, **5**, 506.
6 Saha, N. C., and Mitra, G. D., *J. Chromatog. Sci.*, 1970, **8**, 84.
7 Svoboda, P. A. T., in 'Gas Chromatography 1962.' ed. van Swaay, M., Butterworths, London, 1962, p. 273.
8 Takacs, J., Szita, C., and Tarjan, G., *J. Chromatog.*, 1971, **5/6**, 1.
9 Rohrschneider, L., *J. Chromatog.*, 1966, **22**, 6.
10 Sapina, W. R., and Rose, L. P., *J. Chromatog. Sci.*, 1970, **8**, 214.
11 Reynolds, W. O., *J. Chromatog.*, 1970, **8**, 685.
12 Schmidt, D. E., Szilagyi, P. I. A., and Green, J. P., *J. Chromatog. Sci.*, 1969, **7**, 248.
13 Coupe, N. B., Jones, C. E. R., and Stockwell, P. B., *Chromatographia*, 1973, **6**, 483.
14 Kazyak, L., personal communication.

15   Levy, R. L., *Chromatog. Rev.*, 1966, **8**, 48.
16   Perry, S. G., *Adv. Chromatog.*, 1968, **7**, 221.
17   Brauer, G. M., in 'Thermal Characterisation Techniques', ed. Slade, P. E., jun., and Jenkins, L. T., Marcel Dekker, New York, 1970, p. 41.
18   Jain, N. C., Fontan, C. R., and Kirk, P. L., *J. Forensic Sci. Soc.*, 1965, **5**, No. 2, p. 102.
19   May, R. W., Pearson, E. F., Scothern, M. D., and Porter, J., *Analyst*, 1973, **98**, 364.
20   Deans, D. R., *Chromatographia*, 1968, **1**, 187.
21   Curry, A. S., Read, J. F., and Brown, C., *J. Pharm. Pharmacol.*, 1969, **21**, 224.